Solar Energy Fundamentals

Solar Energy Fundamentals

By Dr. Robert K. McMordie

Edited by
Mitchel C. Brown
Dr. Robert S. Stoughton

THE FAIRMONT PRESS, INC.

CRC Press
Taylor & Francis Group

Library of Congress Cataloging-in-Publication Data

McMordie, Robert K.
  Solar energy fundamentals / by Robert K. McMordie ; edited by Mitchel C. Brown, Robert S. Stoughton.
    p. cm.
  Includes bibliographical references and index.
  ISBN 0-88173-681-3 (alk. paper)
  ISBN 0-88173-682-1 (electronic)
  ISBN 978-1-4665-1400-3 (Taylor & Francis : alk. paper)

1. Solar energy. 2. Solar energy--Equipment and supplies--Mathematical models. I. Brown, Mitchel C. II. Stoughton, Robert S. III. Title.

  TJ810.M18 2012
  621.47--dc23

                                                        2011050734

Published by The Fairmont Press, Inc.
700 Indian Trail
Lilburn, GA 30047
tel: 770-925-9388; fax: 770-381-9865
http://www.fairmontpress.com

Distributed by Taylor & Francis Ltd.
6000 Broken Sound Parkway NW, Suite 300
Boca Raton, FL 33487, USA
E-mail: orders@crcpress.com

Distributed by Taylor & Francis Ltd.
23-25 Blades Court
Deodar Road
London SW15 2NU, UK
E-mail: uk.tandf@thomsonpublishingservices.co.uk

Printed in the United States of America
10 9 8 7 6 5 4 3 2 1

ISBN 0-88173-681-3 (The Fairmont Press, Inc.)
ISBN 978-1-4665-1400-3 (Taylor & Francis Ltd.)

While every effort is made to provide dependable information, the publisher, authors, and editors cannot be held responsible for any errors or omissions.

Dedication

On behalf of our father, we dedicate this book to his dear friend and companion, Judy Lewis. During their time together in Brackettville, Texas, in early 2010, Judy spent many rainy days reading and editing the book. Our dad talked often about how much Judy's support meant to him. Most importantly, she gave our dad unconditional encouragement and love during the last four years of his life. For this, we are forever grateful.

Table of Contents

FOREWORD ix

Chapter   *Page*

1 INTRODUCTION 1

2 INTRODUCTION TO ENERGY TERMS 3

3 THE SUN & THE EARTH... 5

4 SUNLIGHT ON THE EARTH.. 9

5 HEAT TRANSFER... 19
  Introduction 19
  Conduction.. 20
  Convection 22
  Radiation 26
  Oppenheim Radiation Networks............................... 35
  Generalized Network Analysis 39
  Closing Remarks 48

6 SOLAR COLLECTORS... 49

7 ABSORBED SOLAR ENERGY 61

8 SOLAR DOMESTIC HOT WATER SYSTEMS 69

9 SOLAR SPACE HEATING.. 73

10 SOLAR ELECTRIC SYSTEMS 79

11 STIRLING ENGINE & SOLAR POWER SYSTEMS 87

12 PASSIVE SOLAR ENERGY.. 93

13 GREENHOUSE SOLAR COLLECTOR.............................. 99

PROBLEMS 101

REFERENCES 107

APPENDICES 109

A   Carnot Efficiency 111
      Stirling Efficiency.. 115

B   Mathematical Techniques for Solving
      Heat Transfer Problems 118

C   Case Studies.. 163

INDEX... 199

Foreword

SOLAR ENERGY FUNDAMENTALS AND COMPUTATIONS

For readers of this book who are not technically trained it is suggested that you skip Chapter 5, Heat Transfer. Also, you don't need to read the parts that are concerned with developing equations. However, you are encouraged to use the Microsoft Excel programs provided on the CD included with this book. A listing of these programs is given below.

It is important to understand that the analysis presented here yields approximate values for incident solar radiation values, absorbed solar radiation values, cost savings from solar systems, and payback periods for solar systems. The analysis does take into account weather data, the collector performance, attenuation of the solar energy coming through the atmosphere, the angle between the surface of the collector and the solar beam, and the clearness of the atmosphere. However, the calculations are not exact and, of course, the historic weather data may not represent the weather in the future. Therefore, use the information given here as a guide and realize the results are not exact.

The CD includes the following Solar Energy Computations Microsoft Excel worksheets:

1. Computations
    *   Computations Index B
    *   Film Coefficient 1 Btu, 2, 3, 4 & 5
    *   NU Glazing
    *   Pressure drop
2. Flat Plate Collectors
    *   No transparent cover
    *   One transparent cover
    *   Two transparent covers
3. Incident-Absorbed Solar Radiation

- Fixed
- One axis azimuth tracking
- One axis tilt angle tracking
- Two axis tracking
4. Thermal Property Values
   - Gases
   - Liquid metals
   - Saturated Liquids
   - Metallic Solids
   - Nonmetallic Solids
5. Radiation Property Values
   - Metallic
   - Nonmetallic
6. Declination
7. Domestic Hot Water
8. Film Coefficient Calculations
9. Forced flow over a single tube
   - Free convection around a horizontal tube
   - Free convection on a vertical surface
   - Free convection on a horizontal surface
   - Free convection between parallel plates
   - Forced convection inside a pipe
10. Incident radiation
11. Length of Day
12. Metric Unit Conversions
13. Passive Window Analysis
14. Pressure Drop in a Smooth Tube
15. Problem Solutions
    - Solutions to the 17 sample problems in the book
16. Quartic Equation Solution
    - Radiation View Factors
    - Equations for 53 different situations
    - View factor index
17. Solar Space Heating
18. Sun Location
19. Weather Data for 150 different cities nationwide

- Monthly Average Temperatures
- Daytime Average Temperatures
- Daily Average Temperatures
- Percent of Possible Sunshine
- Clearness Number
- Latitude

When the data are a function of temperature the data points are given along with linear interpolation between data points and linear extrapolation outside the data points. Also, for the properties as a function of temperature, a polynomial equation is given along with the accuracy of the equation in matching the data points. These equations can be readily "exported" to Excel workbooks to support heat transfer computations.

Preface

During the last year of our father's life, he worked diligently on his book—*Solar Energy Fundamentals*—with the expectation of delivering a comprehensive understanding of solar energy to the public. He envisioned his book being read by people with varying educational backgrounds and experience, and wanted the information presented in the book to be used in real life projects. To this end, he developed an impressive system of Excel-based companion computations for the reader to readily put the information in the book to practical use.

This book is a compilation of decades of knowledge that spanned Dr. McMordie's career as a mechanical engineer specializing in heat transfer and thermodynamics in the solar and aerospace industries. Dr. McMordie received his Bachelor's and Master's degrees in Mechanical Engineering from the University of Texas. He then obtained his doctorate in Mechanical Engineering from the University of Washington while raising his four young daughters with his wife, Janie. During this time, he worked at Boeing, where he performed hydraulic testing of the Minuteman ground systems and thermal analysis on the Bomarc missile and Dyna-Soar re-entry vehicle.

After leaving Boeing, Dr. McMordie joined the University of Wyoming in Laramie as an associate professor of Mechanical Engineering, teaching courses in heat transfer and thermodynamics. He excelled at teaching and loved bringing the world of engineering to his students. They recognized his leadership by awarding him *Outstanding Engineering Professor of the Year* in 1969.

That same year Dr. McMordie moved with his wife and six daughters (the last two born in Laramie) to Denver, CO, and joined Martin Marietta, working there through his retirement in 1990. At Martin Marietta (now Lockheed Martin) he performed thermal analyses and tests on the Viking Thermal Switch and the Biology Instrument. Dr. McMordie was the lead thermal engineer on the

Space Station Program and on several independent research and development projects.

Dr. McMordie also worked in the field of solar energy while at Martin Marietta including the development of a solar energy design program (Solcost), which is a thermal analysis software program for solar collectors. He was the lead engineer for the design and testing of a 5-megawatt molten salt central receiver at Sandia National Laboratory. The "solar power tower" used sun-tracking mirrors to concentrate solar energy to generate electric power.

Dr. McMordie's love for teaching continued. He went on to instruct courses in heat transfer, thermodynamics and solar energy at the Martin Marietta Institute. He also taught several semesters of the thermal control portion of a spacecraft design course at the University of Colorado.

It is our wish that our father's love of solar engineering be continued through the readers of his book, soaking in his expansive knowledge and applying the information in real world projects.

Acknowledgments

We would like to express our deep gratitude to Mitchel Brown and Dr. Robert Stoughton who spent many hours providing a comprehensive technical review of our father's book after his passing. Mitchel and Robert, his sons-in-law, reviewed the book's text and companion computations and worked with the publisher to provide important information during the production of the book. We also would like to thank James Kidd and Stan Hightower, longtime friends and colleagues of our father, for reviewing and providing comments on the book as well. Without all of their contributions, this book would not have been published.

The daughters of Dr. R.K. McMordie

Chapter 1

Introduction

Solar energy can be used in several different ways to provide people with a readily available power source. Basic and early uses of solar energy include using the sun for common tasks such as drying clothes, curing meat, and heating water. In the past few decades, solar energy has been used in residential and commercial settings to power heating and cooling systems such as air conditioners, furnaces, heating, ventilation and air conditioning (HVAC), and water heaters. A common method of harnessing the sun's energy for these uses is photovoltaics—a process of using solar cells to convert light directly into electricity.

On a more substantial scale, power plants can be built using solar energy as the energy source. These plants consist of a field of sun tracking mirrors, called heliostats, which reflect the sun to a solar receiver mounted on a tower. The tower is 200 or more ft high and the receiver captures the solar energy in a working fluid. The working fluid can be water or some other heat transfer fluid. If the working fluid is water, the solar energy boils the water and the resulting steam is used in a conventional steam power plant. If the working fluid is something other than water, it is used to boil water and, again, the steam is used in a conventional power plant.

This book covers the basics of solar energy. Relevant equations and calculations are shown in the text, programmed using Microsoft Excel. The included CD contains Excel programs with calculations for incident solar radiation calculations on surfaces positioned at any orientation, payback period estimates for home heating, photovoltaic system performance, domestic hot water heating systems, and much more. The CD contains a substantial amount of thermal and radiation property data. Much of this data (thermal conductivity, density, etc.) is temperature dependent and

1

is given in the form of polynomial equations, which are handy for heat transfer calculations. Weather and solar data for over 150 US cities are also included on the CD.

Chapter 2

Introduction to Energy Terms

What are temperature, heat, and power? Temperature is a measure of how hot or cold an object is. Fahrenheit (F) is the temperature scale used for British units while Celsius (C) is the metric temperature scale. Centigrade (C) is the same as Celsius. For the Fahrenheit scale, 32 is the freezing point of water and 212 is the boiling point of water. For the Celsius scale, 0 is the freezing point of water and 100 is the boiling point. The equation which relates the two temperature scales is $F = 1.8 * C + 32$.

There is a limit on how cold an object can become. This limit is -459.67 °F. In the British units, there is a temperature scale based on the minimum low temperature, which is called the Rankine scale. For this temperature scale -459.67 °F = 0 °R (Rankine). The relationship between the Rankine and Fahrenheit scales is °R = °F +459.67. The Rankine scale is called an absolute temperature scale, and this scale is used in radiation heat transfer analysis.

Heat is thermal energy defined as energy in transition due to a temperature difference. For example, if a hot rock is dropped into a pan of cold water, heat will flow from the rock into the water. The temperature of the rock will lower and the water temperature will go up. A unit of heat is the British thermal unit, (Btu). One Btu is the amount of heat required to raise one pound of water one degree Fahrenheit. It takes 3413 Btu's to equal 1 kilowatt- hour.

Power is defined as energy per unit time, for example, Btu's per hour. A kilowatt is also a measure of power and is equal to 3413 Btu/hour. One horsepower is equal to 2544 Btu/hour.

Chapter 3

The Sun & The Earth

Our sun, with a surface temperature of about 10,000°F, sends a constant stream of radiation energy into the space which surrounds it. At the surface of the sun, the thermal radiation is about 20,100,000 Btu/(hr-ft²). The thermal radiation, which arrives at the Earth's outer atmosphere, is about 430 Btu/(hr-ft²) and is reduced, on a clear day, to about 350 Btu/(hr-ft²) after passing through the Earth's atmosphere. On a cloudy day, the solar radiation reaching the Earth's surface is much less than the 350 Btu/(hr-ft²) value.

The diameter of the Earth is about 7900 miles, and the distance around the Earth at the equator is about 24000 miles. The distance from the Earth to the sun is shown in Figure 3-1 and is about the same distance as travelling around the Earth at the equator 3800 times. It takes light, travelling at 186,000 miles per second, over eight minutes to span the distance from the sun to the Earth.

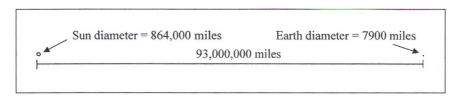

Sun diameter = 864,000 miles   Earth diameter = 7900 miles
93,000,000 miles

Figure 3-1. The Sun and Earth, Approximately to Scale

The Earth rotates around the sun as shown in Figure 3-2. The axis of rotation of the Earth is tilted 23.45° relative to the plane of the Earth's orbit. This tilt causes the seasons that are experienced on Earth.

Figure 3-3 is a view of the Earth at the northern hemisphere winter solstice looking parallel to the Earth's orbit plane. Notice

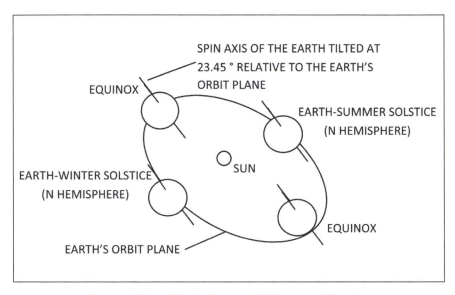

Figure 3-2. Earth's Postions in Orbit around the Sun

that if you are located at a north latitude greater than 66.55° on the winter solstice, you will be in the Earth's shadow (at least a partial shadow) for a 24-hour period. Due to refraction and the fact that the sun is a disk and not a point, you will see the sun or at least a portion of the sun for a part of the day at latitudes greater than 66.55°.

Figure 3-4 is a view of the Earth at the northern hemisphere summer solstice looking parallel to the Earth's orbit plane. Notice that if you are located at a north latitude greater than 66.55° on the summer solstice, you will have, on a clear day, sunshine continuously over a 24-hour period.

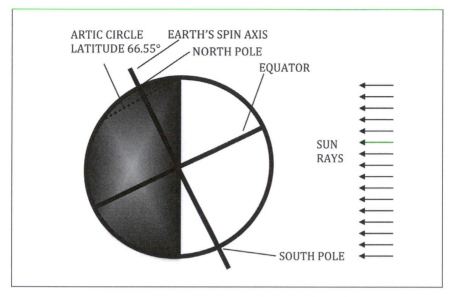

Figure 3-3. Earth at Winter Solstice (N Hemisphere)

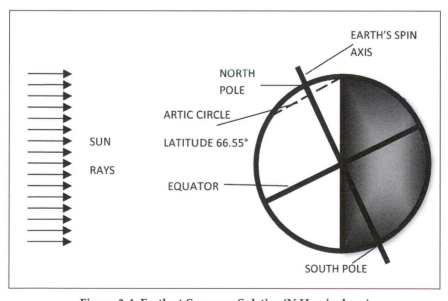

Figure 3-4. Earth at Summer Solstice (N Hemisphere)

Chapter 4

Sunlight on the Earth

In order to estimate the amount of solar energy striking a surface (for example, a solar collector) one must determine the location of the sun relative to the surface at any given time on any given day. This is done by determining the angle θ, which is described in Figure 4-1. The angle θ is defined by Equation (4-1) given below. Equation (4-1) is taken from ref (1).

If the area of the flat surface is A and the intensity of the incoming solar radiation is I_{DN}, then the amount of solar energy (Q) striking the front side of the flat plate on a clear day is:

$$Q = A * I_{DN} * \cos \theta, \text{ if } \theta > 90°, Q = 0 \text{ (from ref 2)} \qquad (4-1)$$

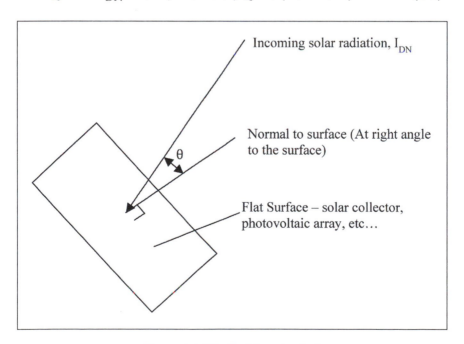

Figure 4-1. The Incident Angle θ

9

For example, if the sunrays are parallel to the surface of the flat plate, $\theta = 90°$ and $\cos \theta = 0$. If $\theta > 90°$ the sun is shining on the backside of the plate and the front side is shadowed from the sun. I_{DN} is defined by Equation (4-7).

$$\cos \theta = \sin \Delta \sin \varphi \cos S—\sin \Delta \cos \varphi \sin S \cos \gamma +$$
$$\cos \Delta \cos \varphi \cos S \cos \omega + \cos \Delta \sin \varphi \sin S \cos \omega \cos \gamma +$$
$$\cos \Delta \sin S \sin \gamma \sin \omega \text{ (from ref 1)} \tag{4-2}$$

Where:

θ = the angle between the normal to the flat surface and the sun beam (solar vector), degrees

φ = latitude, degrees

Δ = declination (the angular position of the sun at solar noon with respect to the plane of the equator) declination is determined from an equation or tables, degrees

S = slope of the collector from the horizontal, degrees

γ = azimuth angle (the angle between the direction the collector face is pointing and south, east of south positive & west of south negative), degrees

ω = hour angle (solar noon is zero and each hour equaling 15° of longitude with mornings positive and afternoons negative), degrees. For example, 14:30 = -37.5°.

An equation for the declination, Δ, is given below (ref. 3). The value varies from +23.45° to –23.45° and the average difference between the actual declination and the calculated declination is 0.0412°. Obviously, the equation is very accurate. To calculate the declination all that is needed is the Day of the Year.

$$\Delta = 180/\pi * (0.006918—0.399912 \cos\beta + 0.070257 \sin\beta - 0.006758 \cos2\beta + 0.000907 \sin2\beta - 0.002697 \cos3\beta + 0.00148 \sin3\beta) \tag{4-3}$$

$$\beta = 2 \pi (DOY - 1)/365 \tag{4-4}$$

$$DOY = \text{Day of the Year}$$

Table 4-1, given below, gives the Day of the Year for each date of the year. To find the declination for February 29, average the declinations for February 28th and March 1st.

Table 4-1. Day of the Year

DAY	JAN	FEB	MAR	APR	MAY	JUN	JUL	AUG	SEP	OCT	NOV	DEC
1	1	32	60	91	121	152	182	213	244	274	305	335
2	2	33	61	92	122	153	183	214	245	275	306	336
3	3	34	62	93	123	154	184	215	246	276	307	337
4	4	35	63	94	124	155	185	216	247	277	308	338
5	5	36	64	95	125	156	186	217	248	278	309	339
6	6	37	65	96	126	157	187	218	249	279	310	340
7	7	38	66	97	127	158	188	219	250	280	311	341
8	8	39	67	98	128	159	189	220	251	281	312	342
9	9	40	68	99	129	160	190	221	252	282	313	343
10	10	41	69	100	130	161	191	222	253	283	314	344
11	11	42	70	101	131	162	192	223	254	284	315	345
12	12	43	71	102	132	163	193	224	255	285	316	346
13	13	44	72	103	133	164	194	225	256	286	317	347
14	14	45	73	104	134	165	195	226	257	287	318	348
15	15	46	74	105	135	166	196	227	258	288	319	349
16	16	47	75	106	136	167	197	228	259	289	320	350
17	17	48	76	107	137	168	198	229	260	290	321	351
18	18	49	77	108	138	169	199	230	261	291	322	352
19	19	50	78	109	139	170	200	231	262	292	323	353
20	20	51	79	110	140	171	201	232	263	293	324	354
21	21	52	80	111	141	172	202	233	264	294	325	355
22	22	53	81	112	142	173	203	234	265	295	326	356
23	23	54	82	113	143	174	204	235	266	296	327	357
24	24	55	83	114	144	175	205	236	267	297	328	358
25	25	56	84	115	145	176	206	237	268	298	329	359
26	26	57	85	116	146	177	207	238	269	299	330	360
27	27	58	86	117	147	178	208	239	270	300	331	361
28	28	59	87	118	148	179	209	240	271	301	332	362
29	29		88	119	149	180	210	241	272	302	333	363
30	30		89	120	150	181	211	242	273	303	334	364
31	31		90		151		212	243		304		365

The equation for hour angle, given in reference 1, is:

$$\omega = (12—TOD) * 15 \tag{4-5}$$

TOD = military time of day in hours and fractions of hours. Example, 2:45 PM is 14.75.

There is no need to worry about having to carry out the calculations given above—they are all contained in the Excel files on the included CD. The file *Sun Location* contains a program to calculate the angle between the normal to a flat plate and the sunbeam, θ. The Excel file *Length of Day* contains a program that computes the length of a given day with sunrise and sun set times. Note that for the computations given in the CD, solar time is used. Solar time is defined as a time with the sun being at its zenith (highest point in the sky) at 12:00 noon.

The equation for the length of day, LOD, is given in reference 1 and is as follows:

$$LOD = (2/15) * cos^{-1} (-tan\varphi\ tan\Delta), hours \tag{4-6}$$

For the computations presented in the CD, it is assumed that the sun is at its zenith at 12 noon for all situations. Therefore, sunrise is at 12—LOD/2 and sunset is at 12 + LOD/2. In equation (6), φ is latitude and Δ is declination.

After establishing where the sun is located relative to a flat surface we need to determine the strength of the sun's radiation at the Earth's surface. This is accomplished by using material from *The 1999 ASHRAE Applications Handbook of the American Society of Heating, Refrigeration, and Air-Conditioning Engineers*. The information given in the ASHRAE Applications Handbook and the cos θ (from above) lets us calculate the clear day strength of the solar radiation striking any given surface at any time during the day.

There are several quantities that are required in order to estimate the intensity of solar radiation on the Earth's surface. Table 4-2 lists these items: A-the apparent solar irradiation at air mass zero, B-atmospheric extinction coefficient, C- indirect coefficient and declination, Δ. Note that air mass zero refers to solar radiation at the Earth, above the Earth's atmosphere.

Notice that the dates used in Table 4-2 are on the 21st of the months. The 21st is used to coincide with equinox and solstice dates. Solar calculations are often linked with monthly average solar and weather data. This means that it is necessary to either move the data in Table 4-2 to mid-month values or to move the monthly average solar and weather data to the 21st of each month. For the computations in the CD, the mid month values are used. Table 4-3 is a repeat of Table 4-2 except the data are interpolated for mid-month values.

Table 4-2. Solar Coefficients at the 21st of each Month

DATE	A	B	C	δ
Jan 21	390	0.142	0.058	-19.9
Feb 21	385	0.144	0.06	-10.6
Mar 21	376	0.156	0.071	0
Apr 21	360	0.18	0.097	11.9
May 21	350	0.196	0.121	20.3
Jun 21	345	0.205	0.134	23.45
Jul 21	344	0.207	0.136	20.5
Aug 21	351	0.201	0.122	12.1
Sep 21	365	0.177	0.092	0
Oct 21	378	0.16	0.073	-10.7
Nov 21	387	0.149	0.063	-19.9
Dec 21	391	0.142	0.057	-23.45

A = Apparent Solar Irradiation at Zero Air Mass

B = Atmospheric Extinction Coefficient

C = Indirect Coefficient
δ = Declination

Table 4-3. Solar Coefficients at Mid-month

DATE	A	B	C	δ
Jan	390.2	0.142	0.058	-21.3
Feb	386.1	0.144	0.06	-13.3
Mar	377.8	0.154	0.069	-2.4
Apr	363.1	0.175	0.092	9.5
May	351.8	0.193	0.117	18.7
Jun	346	0.203	0.131	23.3
Jul	344.2	0.207	0.136	21.7
Aug	349.8	0.202	0.124	14.3
Sep	362.3	0.182	0.098	3.3
Oct	375.6	0.163	0.076	-8.2
Nov	385	0.151	0.065	-18.3
Dec	390.1	0.144	0.058	-23.2

A = Apparent Solar Irradiation at Air Mass Zero

B = Atmospheric Extinction Coefficient

C = Indirect Coefficient
δ = Declination

In Table 4-3 the data are given on a monthly basis. This is because the approach used to estimate yearly performance of the solar systems provided on the CD is based on monthly average data. This includes monthly average weather data (average temperatures) and monthly average solar data (Percent of Possible Sunshine) along with the data from Table 4-3. Calculations are performed at midmonth to obtain average daily values. These values are multiplied by the number of days of the month to establish monthly values. The monthly values are summed to obtain yearly estimates. The basic approach used here is similar to the method used by the Solcost program, a solar program the author was involved in while working for Martin Marietta (now Lockheed Martin).

The equation that defines the intensity of the sun at the surface of the Earth on a clear day is:

$$I_{DN} = Ae^{\frac{-B}{\sin(\lambda)}} \tag{4-7}$$

(The subscript DN refers to direct normal. In other words I_{DN} is the intensity of the direct solar rays on a surface normal to the solar vector.)

Where:

A	=	apparent solar irradiation at air mass zero (Table 4-3)
B	=	atmospheric extinction factor (Table 4-3)
λ	=	solar altitude angle

The solar altitude angle is given by:

$$\text{Sin } \lambda = \sin \Delta \sin \varphi + \cos \Delta \cos \varphi \cos \omega \tag{4-8}$$

Where:

Δ	=	declination
φ	=	latitude
ω	=	hour angle (Equation 5)

Figure 4-6 illustrates the geometry associated with I_{DN}.

Along with the direct normal solar radiation, there is an indirect component of solar energy which will strike a surface. This indirect component is diffused or scattered solar energy which comes from the sky. For example, a surface will receive indirect solar radiation even if it is in the shade. The indirect component, I_{IND}, is defined by the following equation:

$$I_{IND} = C \, I_{DN} \, F_{SS} \tag{4-9}$$

Where:

C	=	indirect coefficient (Table 4-3)
I_{DN}	=	direct normal solar radiation (Equation 4-7)
F_{SS}	=	radiation view factor between surface and the sky

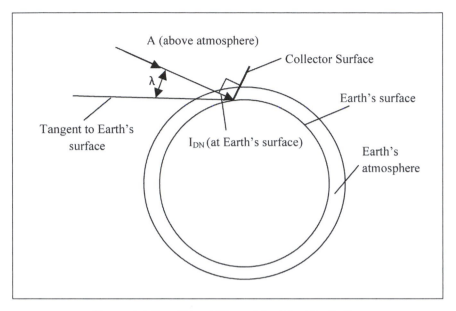

Figure 4-6. I_{DN}, Direct Normal Incident Radiation

$$F_{SS} = 0.5 (1 + \cos S)$$
$$S = \text{tilt angle of the surface}$$

The angle S is shown in Figure 4-7.

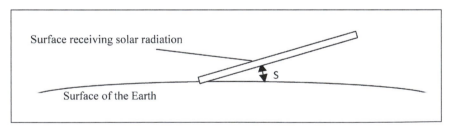

Figure 4-7. The Surface Angle S

An estimate of the total solar energy incident on a surface on the surface of the Earth is as follows:

$$I = I_{DN} [C_N \cos \theta + C/C_N * 0.5 (1 + \cos S)] \qquad (4\text{-}10)$$
$$(\text{If } \theta > 90°, \text{ set } \cos \theta = 0)$$

Where:

I = total solar radiation energy on a surface for an average clear day

I_{DN} = direct normal solar radiation (Equation 4-7)

C_N = clearness number

C = indirect coefficient (Table 4-3)

S = tilt angle of the surface

Clearness numbers are given in Figure 4-4 and are from work by Threkeld and Jordan, ref (2). Clearness numbers account for differences in the atmosphere between regions that are high and dry and low elevations that have high humidity. The clearness number is multiplied by the direct component of the solar radiation but divided into the indirect component. The rationale for this is that on a very clear day the direct solar radiation will be a little higher than on an average day. If the direct solar radiation is a little high, the indirect solar radiation will be a little less than an average day.

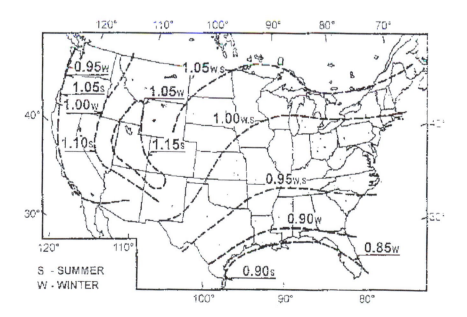

Figure 4-4. Clearness Numbers for the United States

I_{DN} and I_{IND} are radiation fluxes (energy per unit time per unit area) at a given time, a given latitude, and a given surface orientation. These are useful quantities, but a more useful quantity is the total solar energy per unit area (Btu/ft^2) over the daylight hours. We can obtain the total daily values of I_{DN} and I by integrating these quantities with respect to time from sunup to sundown. If we integrate I from sunup to sundown, we have the average daily incident clear day solar radiation on a surface. If this number is multiplied by the percent of possible sunshine for the surface location, we obtain an estimate of the actual solar energy striking the surface per day. This daily average is multiplied by the number of days of the month to obtain monthly averages. We repeat the calculations for each month of the year, sum these values, and obtain an estimate of the yearly radiation. Percent of possible sunshine, as the name implies, is the percent of daylight hours when the sun shines. In the file *Weather Data-F*, there are listed values of percent of possible sunshine and clearness numbers for 150 US cities.

We will conclude this current section with a sample problem. The problem is to determine an estimate of the incident solar radiation on a ten square foot flat surface. The surface is tilted 30° from horizontal, is pointing due south (azimuth angle = 0), and is located in Denver, Colorado.

The first step is to open the Microsoft Excel file entitled *Incident Radiation*. Next open the Excel file, *Weather Data-F*. In the *Weather Data-F* file go to the data for Denver, Colorado. Note the cities in the *Weather Data-F* file are first in alphabetical order by state and then in order by city. Highlight the Denver data starting with the cell labeled Denver and moving to the right and down to include the cell directly under the clearness number for December cell. The highlighted field includes all the data for Denver. Copy the Denver data and open the *Incident Radiation* file. Highlight the cell (A8) which contains the name of the city, it might be any city, and paste. You should now have the Denver data in the *Incident Radiation* file. Now enter the tilt angle in cell B4 and the azimuth angle in cell B5. The yearly average radiation is given in cell B16 and is 8.010E+05 $Btu/(ft^2$-year). The yearly average on the ten square foot surface is 8.010E+06 Btu/year.

Chapter 5

Heat Transfer

INTRODUCTION

Before embarking on this chapter, the reader should be aware that this chapter contains complex technical information. It is recommended that readers who are not technically inclined skip this chapter. An understanding of the fundamentals of solar energy can be acquired without the need to master the material in this chapter.

In the preceding section, a method for estimating the solar radiation incident on a surface was developed. Next, we need to determine the solar energy *absorbed* by a surface. Also, we need to be able to calculate heat losses from structures so that, for example, we can estimate the required size of a solar energy system. These needs require an understanding of heat transfer.

The field of heat transfer is exciting and challenging. There are heat transfer problems in the fields of solar energy, aerospace, automotive engineering, and home building, to name a few. Disciplines used in heat transfer are mathematics, numerical analysis, testing, and instrumentation. Among the early contributors to the field of heat transfer were Newton, Fourier, Plank, and Boltzmann.

Heat transfer is divided into three major areas: conduction, convection, and radiation. Conduction is associated with thermal energy transfer through matter in the absence of fluid motion. Convection is concerned with thermal energy transfer between a flowing fluid and a solid interface. Radiation is energy transfer via electromagnetic waves.

CONDUCTION

The fundamental equation for steady-state, one dimensional*
heat conduction in rectangular coordinates is:

$$Q = \left(\frac{kA}{\Delta x}\right)(T_1 - T_2)$$ (5-1)

In Equation (5-1), Q represents the energy transfer rate (Btu/
hr), T the temperature (°F), k the thermal conductivity (Btu/(hr-
ft-°F)), Δx, the length of the heat transfer path (ft), and A the area
normal to the heat transfer direction (ft²). The elements of Equation
(5-1) are illustrated in Figure 5-1. In this figure there is a wall with
a temperature difference imposed across the wall from inside to
outside. Think of a wall in a house in the winter. The temperature
on the inside of the wall is high relative to the temperature on the
outside. This temperature difference causes a heat flow through
the wall from inside to outside. Inspecting Equation (5-1), heat
transfer through the wall is directly proportional to the size or area
(A) of the wall and to the temperature difference $(T_1 - T_2)$ across
the wall. The larger the area (A) and the temperature difference
between the two surfaces $(T_1 - T_2)$ the larger the heat transfer. Also,
heat transfer (Q) is inversely proportional to the wall thickness.
In other words, if the wall is made thicker, Δx increases, and then
the heat transfer is decreased. So far, the capability of the wall's
material to transfer heat has not been accounted for. For example,
if the wall is made of aluminum it will transfer more heat than a
wooden wall if the wall thickness, size, and temperature difference across the wall are the same for both cases. The heat transfer
capability of the wall material is accounted for by the coefficient
of thermal conductivity, k.

Figure 5-1 illustrates steady-state—one dimensional conduction in rectangular coordinates—while Figure 5-2 illustrates the

---

*Steady state refers to a condition where all variables do not change with time.
One-dimensional refers to heat transfer moving in only one direction.

same in spherical and cylindrical coordinates.

The steady-state conduction equation in infinitesimal form is given by:

$$Q = - kA \ \frac{dT}{dx}$$

(5-2)

In Equation (5-2), Q is the heat transfer rate (Btu/hr), k is the thermal conductivity (Btu/(hr-ft-°F)), A is the heat transfer area normal to the direction of heat flow (ft^2), and dT/dx is the tempera-

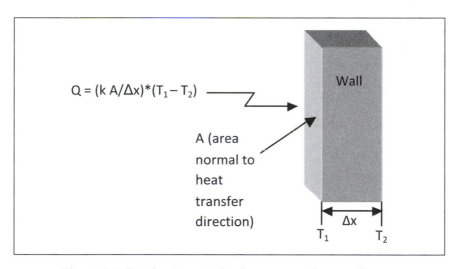

Figure 5-1. Steady-state conduction, rectangular coordinates

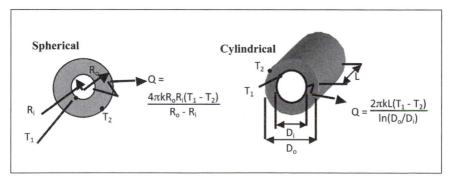

Figure 5-2. Steady-State Conduction, Spherical and Cylindrical Coordinates

ture gradient along the direction that heat flows (°F / ft). Heat flows from higher to lower temperatures so a negative value of dT / dx causes the heat to flow in the positive x direction. Therefore, dT / dx must be negative to account for heat flowing in the positive x direction.

CONVECTION

Convection is thermal energy transfer between a flowing fluid and a surface. For example, when you cool from a fan blowing air across your skin, this is convection heat transfer. There are two major types of convective heat transfer: forced convection and free (or natural) convection. In forced convection, the fluid flow is *forced* by a pump, a fan, the wind etc. If the solid surface and the fluid are at different temperatures, there will be heat transfer between the surface and the fluid. Free convection occurs in the absence of a mechanism to force the fluid flow—the flow itself is caused by the temperature difference between the fluid and the solid surface. In order to understand free convection, consider a solid surface and an adjacent fluid that is at rest, and the surface and the fluid are at the same temperature. Since the fluid and surface are at the same temperature, there is no heat transfer between the surface and the adjacent fluid. Now increase the temperature of the surface so that there is a temperature difference between the surface and the fluid. This temperature difference will increase the temperature of the layer of fluid next to the solid surface. This fluid layer, being at a higher temperature than the bulk of the fluid further away from the solid surface, will also have a lower density than the bulk fluid away from the surface. This density difference will cause the fluid adjacent to the wall to rise due to buoyancy effects. For free convection, the surrounding fluid is essentially at rest with only the fluid adjacent to the solid surface circulating.

The basic convection equation was developed by Newton and is given as:

$$Q = h \, A(T_a - T_w) \tag{5-3}$$

In Equation (5-3), Q is the heat transfer (Btu/hr), h is the convection film coefficient (Btu/(hr-ft^2-°F)), A is the area (ft^2), and T_a and T_w are the temperatures of the ambient fluid and the solid surface, respectively (°F). Small samples of various types of convection situations are illustrated in Figure 5-3.

In Figure 5-3, the terms laminar flow and turbulent flow refer to the basic characteristic of fluid flow. Laminar flow exists in a tube when the fluid flow rate is below a given value. In laminar flow in a straight tube, the fluid particles all move in a straight line. As the flow rate increases, there will come a point where the fluid transitions from laminar to turbulent flow. The fluid particles in turbulent flow in a straight tube do not move in a straight line. The fluid particles "bounce" around as they travel generally down the tube. The Reynolds number, which is dimensionless, gives a ratio of inertial forces to viscous forces in the fluid. The transition from

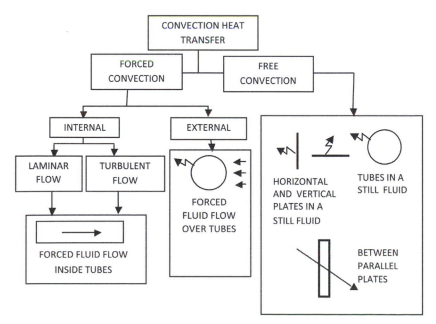

Figure 5-3. Convection Heat Transfer Situations

laminar to turbulent flow in a circular tube occurs at a Reynolds number of approximately 2300. Above this value, inertial forces dominate and the flow is turbulent, while below this value, viscous forces dominate and the flow is laminar. The Reynold's number is given by

$$Re = \rho VD / \mu \tag{5-4}$$

In Equation (5-4) ρ is the fluid density (lb_m / ft^3), V is the average velocity of the fluid (ft/sec), D is the inside diameter of the tube (ft), and μ is the fluid viscosity ($lb_m / (ft\text{-}sec)$).

There are several dimensionless numbers that are important in the field of convection heat transfer. The Nusselt number gives a ratio of convective to conductive heat transfer across the fluid boundary layer. The Prandtl number gives a ratio of the viscous diffusion rate to the thermal diffusion rate, and the Grashof number gives a ratio of buoyancy to viscous forces in a fluid. The equations which define these numbers are as follows:

$$Nu = \text{Nusselt number} = h\,D / k \tag{5-5}$$
$$Pr = \text{Prandtl number} = \mu\,C_p / k \tag{5-6}$$
$$Gr = \text{Grashof number} = \rho^2\,g\,\beta\,\Delta T\,D^3 / \mu^2 \tag{5-7}$$

Where,

h   = convection film coefficient, $Btu / (hr\text{-}ft^2\text{-}°F)$

D   = characteristic length, ft (inside diameter for flow inside a tube, outside diameter for convection off a tube surface, the height of a vertical plate, etc.)

k   = thermal conductivity of the fluid, $Btu / (hr\text{-}ft\text{-}°F)$

μ = viscosity of the fluid, $Lb_m / (ft\text{-}hr)$

Cp   = specific heat of the fluid, $Btu / (lb_m\text{-}°F)$

ρ = density, Lb_m / ft^3

g   = acceleration of gravity = 32.174 ft/sec² (at sea level on the Earth's surface)

β = coefficient of volume expansion, $1/K = -\dfrac{1}{\rho} = \left(\dfrac{\partial \rho}{\partial T}\right)_P$ = $1/T$ for an ideal gas

T = temperature in degrees Rankine, $°R = °F + 459.67$

ΔT = temperature difference between the bulk fluid (the fluid away from the solid surface) and the solid surface, $°F$

In general, property values are evaluated at the average fluid temperature. With regard to free convection, property values are based on the film temperature, which is the average of the surface temperature and fluid temperature away from the surface (bulk fluid temperature). The exception to this is the coefficient of volume expansion, which is evaluated at the bulk fluid temperature.

An example of the equations used to evaluate film coefficients follows. The equations presented here are for forced convection in a tube. Solutions of these equations along with film coefficient computations for the remaining situations shown in Figure 5-3 are given in the computations folder in the included CD.

The average Nusselt number for laminar flow (Reynolds number <2300) in a tube with a constant wall temperature is given as:

$$Nu = \frac{hD}{k} = 3.66 + \frac{0.0668(D/L)RePr}{1 + 0.04[(D/L)RePr]^{2/3}} \tag{5-8}$$

Equation (5-9) represents the average Nusselt number for laminar flow (Reynolds number <2300) in a tube with a uniform heat input along the tube length. Uniform heat input can be approximated by electrically heating the tube.

$$Nu = \frac{hD}{k} = 4.36 + \frac{0.0668(D/L)RePr}{1 + 0.04[(D/L)RePr]^{2/3}} \tag{5-9}$$

In Equations (5-8) and (5-9), D is the inside diameter of the tube, L is the length of the tube, Re is the Reynolds number, and Pr is the Prandtl number. These equations are from reference 4. The film coefficients resulting from Equations (5-8) and (5-9) are average values over the tube length.

For turbulent flow (Reynolds Number >2300), see Equation (5-10) and reference 2, page 219:

$$Nu = \frac{hD}{k} = 0.023 \, Re^{0.8}Pr^N \tag{5-10}$$

N = 0.4 for heating
N = 0.3 for cooling

Use the average, or bulk, temperature of the fluid to evaluate the fluid properties for the equations given above for forced flow in tubes. Often when analyzing forced convection problems for flow in a tube, the fluid velocity is not given directly. Many times the fluid flow rate will be given either in volume rate of flow (ft^3/sec) or mass rate of flow (lb_m/sec). Volume rate of flow (ft^3/sec) equals average fluid velocity (ft/sec) times cross sectional area of tube (ft^2). Mass rate of flow (lb_m/sec) equals fluid density (lb_m/ft^3) times average fluid velocity (ft/sec) times cross sectional area of tube (ft^2).

RADIATION

All matter radiates electromagnetic energy in the form of "small bundles of energy" called photons. Photons travel at the speed of light and have zero mass. The photons are emitted from the surfaces of all matter and the general equation representing the total energy per unit time per unit surface area, q (Btu/(hr-ft^2)), is given by:

$$q = \varepsilon \sigma T^4 \tag{5-11}$$

Equation (5-11) was first enunciated by Stefan who established the relationship experimentally. Later, Boltzmann developed the equation theoretically. It is interesting to note that the numerical value of the constant σ (known as the Stefan-Boltzmann constant) defined by Stefan in 1879 is very close to the currently accepted value of 0.1714 x 10^{-8} Btu / (hr-ft^2-°R^4). The remaining terms in Equation (5-11) are the emissivity, ε, (a dimensionless number), and T, the absolute temperature in °R (Rankine). Emissivity (or emittance) is a dimensionless thermophysical property that is a measure of a surface's ability to emit radiation energy. Emissivity values vary between zero and one. A perfect emitter has an emissivity of one and is called a black body. Certain black paints have emissivities very close to one. Polished gold or silver surfaces have emissivities in the neighborhood of 0.05.

Surfaces not only emit photons, they also absorb photons. Consider an enclosure that is maintained at a uniform temperature, and inside the enclosure a perfect vacuum exists. Due to the vacuum, there will be no conduction or convection in the enclosure, and radiation will be considered the only mode of heat transfer present. All the interior surfaces of the enclosure and all objects within the enclosure will emit radiation according to Equation (5-11). For the objects within the enclosure to remain at a uniform temperature, which we know will indeed happen, there must be an absorption of radiant energy (photons) to balance the emission. This is the case and it can be proven that at a given wavelength (radiation has a wave characteristic as well as a particle characteristic), the emissivity is identically equal to the absorptivity (Kirchoff's Law). Absorptivity is defined as the percentage of incident radiant energy that is absorbed by a surface.

Surface radiation properties, including absorptivity, are described in Figure 5-4. In this figure a transparent plate is irradiated by an incoming radiation source, G. All of the incoming radiation will be reflected, absorbed, or transmitted. The percentage of incoming radiation that is reflected is defined as the reflectivity, the percentage absorbed is defined as absorptivity, and the percentage transmitted is defined as transmissivity.

If a surface is opaque, the transmissivity is zero and $\alpha = 1 - \rho$.

Let us focus again on the radiant energy emitted from a surface. The equation that defines the monochromatic radiation

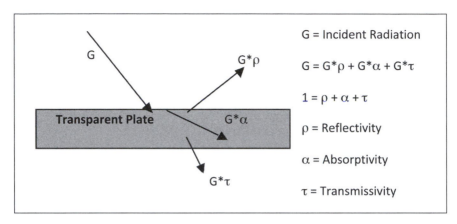

Figure 5-4. Radiation Surface Properties

energy (radiation energy at a given wavelength), $E_{b\lambda}$ emitted from the surface of a black body (emissivity = 1) is given by Planck's equation:

$$E_{b\lambda} = \frac{C_1 \lambda^{-5}}{\left(\dfrac{C_2}{e^{\lambda T}} - 1 \right)} \qquad (5\text{-}12)$$

In Equation (5-12), $E_{b\lambda}$ is the monochromatic emissive power, (emissive power at a given wavelength or energy per unit time per unit area per unit wave length), C_1 and C_2 are constants, λ is the wave length, and T is the absolute temperature. For a non-black body, the monochromatic emissivity (emissivity at a given wavelength), λ, must be included in Planck's equation. To determine the total emission from a surface the product of Equation (5-12) and the surface's monochromatic emissivity must be integrated between 0 and ∞, as shown below.

$$q = \int_0^\infty \frac{\varepsilon_\lambda C_1 \lambda^{-5}}{e^{\frac{C_2}{\lambda T}} - 1} \, d\lambda \qquad (5\text{-}13)$$

If ε_λ is a constant, independent of wavelength, Equation (5-13) reduces to Equation (5-11), the Stefan-Boltzmann equation.

Equation (5-13) was used to generate the curves shown in Figure 5-5. The area beneath each curve represents the total black body (emissivity = 1) emissive power in Btu/(hr-ft^2). An important characteristic of the curves shown in Figure 5-5 is that the curves shift to the right, to longer wavelengths, as the temperature decreases. At the temperature of the sun, approximately 10,000°R, the radiation emission is centered at 0.5 microns, while at room temperature, say 540°R, the radiation emission is centered at 9.7 microns. This is shown in Figure 5-6, which gives the monochromatic emissive power of the sun at the surface of the Earth and the monochromatic emissive power of a surface at room temperature, 540°R.

The difference between the curves labeled "sun" in Figure 5-5 and labeled "solar" in Figure 14 is that the curve from Figure 5-5 shows the emission at the sun's surface, and the curve from Figure 5-6 shows the sun's emission at the Earth's surface. The difference in the two values is the square of the radius of the sun divided by the square of the distance from the sun to Earth.

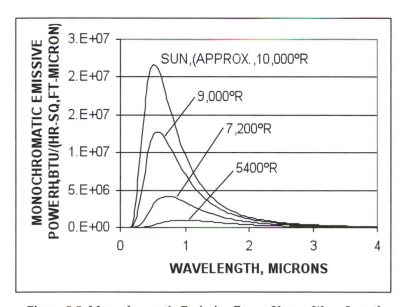

Figure 5-5. Monochromatic Emissive Power Versus Wave Length

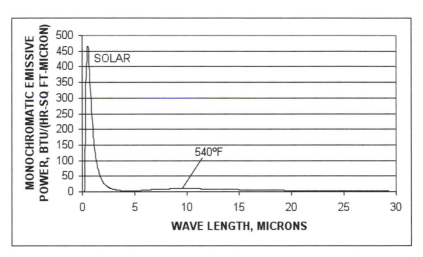

Figure 5-6. Monochromatic Emissive Power Versus Wave Length for Solar Energy at the Earth's Surface and for a Surface at Room Temperature (540°R)

The emissivity of white paint versus wavelength is the solid line shown in Figure 5-7. The curve is also the absorptivity versus wavelength because monochromatic emissivity is identically equal to monochromatic absorptivity. The dotted lines on Figure 5-7 represent the IR emissivity, ε_{IR}, and the solar absorptivity, α_S, of the white paint. IR emissivity is the average emissivity over the IR wave band, approximately 3 to 70 microns. Solar absorptivity is the average absorptivity in the solar wave band, approximately 0 to 3 microns. Equations which define the radiation properties, ε_{IR} and α_S, follow.

$$\varepsilon_{IR} = \frac{\int_{\infty}^{3} \frac{\varepsilon_\lambda C_1 \lambda^{-5}}{e^{\frac{C_2}{\lambda T}} - 1} d\lambda}{\sigma T^4} \tag{5-14}$$

In Equation (5-14) ε_λ is the monochromatic emissivity, C_1 and C_2 are constants, λ is the wavelength, T is the absolute temperature, and σ is the Stefan-Boltzmann constant. Note that IR emissivity is a weak function of surface temperature, and generally the tempera-

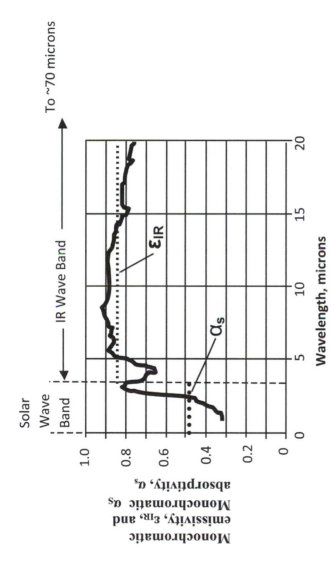

Figure 5-7. Monochromatic Emissivity, Monochromatic Absorptivity, Solar Absorptivity, and Infrared (IR) Emissivity of White Paint Versus Wave Length.*

*White Paint Properties Courtesy NASA; Ref. 6. Volume 1, Page 149

ture must be a hundred degrees different from room temperature
to show a significant change from the room temperature value.

$$\alpha_S = \frac{\int\limits_{3}^{0} \alpha_\lambda S_\lambda d\lambda}{\sigma T^4} \tag{5-15}$$

In Equation (5-15), α_λ is the monochromatic absorptivity, S_λ the
monochromatic solar flux at the Earth's surface, λ is wavelength, σ
is the Stefan-Boltzmann constant, and T is temperature. There are
a large number of values of solar absorptivity and IR emissivity
given in the included CD. These values are found in the file *Radiation Properties* located in the *Property Values* folder.

In order to illustrate the use of ε_{IR} and α_S, consider a surface
that is perfectly insulated on the backside, placed in space, and
irradiated with solar energy normal to the surface. The energy
balance for this situation is as follows:

$$Q_{absorbed} = q_{emitted} \tag{5-16}$$

$$G_s \alpha_s = e_{IR} \sigma T^4 \tag{5-17}$$

solving for the temperature T

$$T = \left(\frac{G_s \alpha_s}{\varepsilon_{IR} \sigma}\right)^{\frac{1}{4}} \tag{5-18}$$

Assume that the solar absorptivity = 0.15 and the IR emissivity = 0.8.

$$T = \left[\frac{(434.9)(0.15)}{(0.8)(0.1714 \times 10^{-8})}\right]^{\frac{1}{4}} = 467°R = 7°F \tag{5-19}$$

In this example, G_s is the solar flux with a value of 434.9 Btu/
(hr-ft^2), α_s is the solar absorptivity of a high quality solar reflector, ε_{IR} is the infrared emissivity of the solar reflector, and σ is the
Stefan-Boltzmann constant, 0.1714 x 10^{-8} Btu/(hr-ft^2-R^4).

The expressions for the net radiation exchange (Q_{net}) between two surfaces that are perfect emitters ($\varepsilon = 1$) and perfect absorbers ($\alpha = 1$), i.e., black surfaces, are given by:

$$Q_{Net1-2} = A_1 F_{1-2} \sigma (T_1^4 - T_2^4) \tag{5-20}$$

$$Q_{Net1-2} = A_1 F_{2-1} \sigma (T_2^4 - T_1^4) \tag{5-21}$$

In Equations (5-21) and (5-22), the absolute temperatures of the surfaces are T_1 and T_2, σ is the Stefan-Boltzmann constant, A is area, and the F terms are view factors, also known as shape factors or configuration factors. A radiation view factor from surface 1 to surface 2, F_{1-2}, is that fraction of the radiant energy directly incident on the receiving surface 2 relative to the total radiant energy leaving the sending surface 1. For example, if the view factor $F_{1-2} = 0.5$, this means that one half of the radiant energy leaving surface 1 will strike surface 2 via a direct path from surface 1 to surface 2. The term "direct path" means that radiation from surface 1 that arrives at surface 2 due to reflections off other surfaces is not included in the view factor evaluation. In Figure 5-8, the quantities involved in the view factor equation are defined.

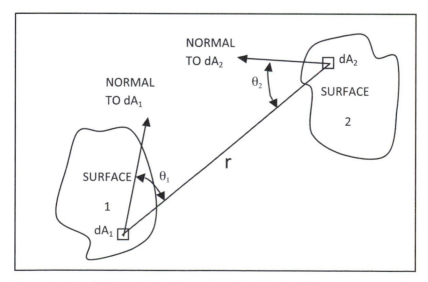

Figure 5-8. Definition of the Quantities Used in the View Factor Equation

The view factor equation from surface 1 to 2 is as follows:

$$F_{1-2} = \frac{1}{A_1} \int_{A_1} \int_{A_2} \frac{\cos\theta 1 * \cos\theta 2 * dA_1 * dA_2}{\pi * r^2} \qquad (5\text{-}22)$$

The view factor equation from surface 2 to 1 is similar:

$$F_{2-1} = \frac{1}{A_2} \int_{A_1} \int_{A_2} \frac{\cos\theta 1 * \cos\theta 2 * dA_1 * dA_2}{\pi * r^2} \qquad (5\text{-}23)$$

View factors, in general, are extremely difficult to evaluate except for the simplest geometry. Notice that the view factor equation contains two area integrals, which means a quadruple integral must be solved to evaluate a view factor. Fortunately, there are existing solutions for many situations. Several solutions are given in the folder entitled *View Factors* on the included CD. The utility of the view factor data given in the Computations folder can be greatly expanded using view factor algebra. Two fundamental rules of view factor algebra are:

1) The sum of all the view factors from a surface must equal unity and

2) The reciprocity rule, $A_x F_{x-y} = A_y F_{y-x}$.

Rule 1 follows from the fact that a view factor is the ratio of the energy leaving an emitting surface and striking a viewed surface to the total energy leaving the emitting surface. The energy striking all the surfaces that the emitting surface sees is equal to the total energy leaving the emitting surface. Therefore, the sum of all the view factors for a given surface must equal unity. A comparison of Equations (5-22) and (5-23) provides the basis for the reciprocity rule. If A_1 in Equation (5-22) is taken to the left side of the equation and A_2 is taken to the left side of Equation (5-23), then the right sides of the two equations are identical. Therefore, $A_1F_{1-2} = A_2F_{2-1}$.

To illustrate the utility of these rules, consider concentric spheres. The inner sphere is defined as surface 1 with area, A_1, and the outer sphere surface 2 with area A_2. Since surface 1 only sees surface 2 then F_{1-2} must equal 1 since F_{1-2} is the sum of all the view factors from surface 1. Using the reciprocity rule, $A_1F_{1-2} = A_2F_{2-1}$, or $F_{2-1} = A_1/A_2 = D_1^2/D_2^2$, where D_1 and D_2 are the diameters of the respective spheres. Surface 2 sees surface 1 but also sees itself. Therefore, we write $F_{2-1} + F_{2-2} = 1$, or $F_{2-2} = 1 - F_{2-1} = 1 - D_1^2/D_2^2$.

OPPENHEIM RADIATION NETWORKS

Equation (5-20) provides a means of calculating radiant energy exchange between "black" surfaces. In Figure 5-9 an approach, developed by Oppenheim[7], is presented that can be used to determine the radiation exchange between non-black surfaces. The restrictions for this approach are as follows:

1. All surfaces are "gray," i.e., all surface emissivities are constant over the wavelength bands applicable to the surfaces temperatures.

2. All surfaces emit and reflect diffusely. (Most surfaces can be considered diffuse except for shiny metallic surfaces.)

3. All surfaces are isothermal (at a constant and uniform temperature.)

4. The radiosities are uniform across each surface. Radiosity is defined as the total radiant energy striking a surface including emitted, reflected, re-reflected, etc. (For most radiation problems, it is reasonable to assume uniform radiosity.)

The characteristics of the Oppenheim network are such that at each given surface, say surface x, the adjacent conductor is of

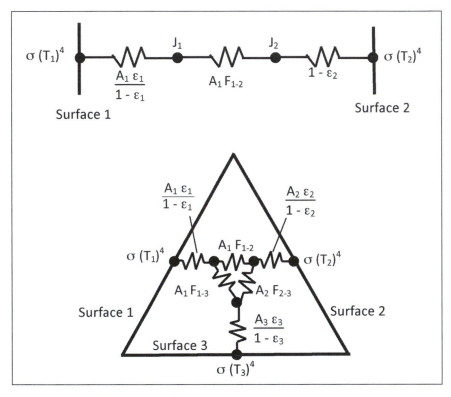

Figure 5-9. Oppenheim Radiation Networks.

the form $A_x \varepsilon_x/(1-\varepsilon_x)$, where A is area and ε is emissivity of the surface. This conductor is connected between the surface node and the dummy J_x node. A surface node is a point which represents a surface and is a junction for two or more conductors. A dummy node does not represent a surface or anything physical—it is merely a junction of two or more conductors. The potential at the surface node is $\sigma(T_x)^4$ and $\sigma(T_J)^4$ at the J node. Connecting each J node to every other J node completes the network. The conductors connecting the J nodes are of the form $A_x F_{x-y}$, where A is area and F the view factor. If a given surface is perfectly insulated or if the emissivity of the node is one (a black surface), the conductor, $A_x \varepsilon_x/(1-\varepsilon_x)$, is eliminated. The effect is, therefore, that the surface potential, $\sigma(T_x)^4$, moves to the J node location.

The Oppenheim approach for analyzing radiation problems

is an extremely useful tool. An example problem using this technique is illustrated in Figure 5-10. The first step in the solution is to determine the necessary view factors. Using the view factor 1 file on the CD, the view factor from the radiator to the wall is found to be 0.29. Because the sum of all view factors from the radiator must equal unity, the view factor from the radiator to space equals $(1-F_{R-W}) = 1 - 0.29 = 0.71$. The view factor from the wall to space is found by first using reciprocity to find the view factor from the wall to the radiator, $A_R \, F_{R-W} = A_W \, F_{W-R}$. Since $A_W = A_R/2$ this results in a value of 0.145 for F_{W-R}. All the view factors from the wall must sum to unity, therefore, the view factor from the wall to space, F_{W-S}, equals $(1-F_{W-R}) = 1-0.145 = 0.855$.

The next step in the solution is to create the Oppenheim network. The network on the lower left-hand side of Figure 5-10 is a basic representation for the problem with $A\varepsilon/(1 - \varepsilon)$ type conductors connecting the surface nodes (with $\sigma \, T^4$ potentials) to the J nodes. The J nodes are connected to each other with conductors made up of the area times the view factor. The network on the left of Figure 5-10 can be immediately reduced to the network shown on the right side of the figure by noting the following characteristics of Oppenheim networks. First, the net heat flow to a surface is given by the conductor value at the surface node $[A\varepsilon/(1 - \varepsilon)]$ times the adjoining potential difference $T^4 - J$. If a surface is perfectly insulated there can be no net heat transfer to the surface, therefore, for this situation $\sigma \, T^4 = J$ because the conductor $[A\varepsilon/(1 - \varepsilon)]$ is a finite number. This causes the elimination of the conductor $[A_W\varepsilon_W/(1 - \varepsilon_W)]$. If a surface is "black" with an emissivity of one, the conductor adjacent to the surface goes to infinity. This has the effect of shorting the conductor and causes the node surface potential ($\sigma \, T^4$) to be one-and-the-same with the adjacent J node. This allows the elimination of the $A_s \, \varepsilon_s/(1 - \varepsilon_s)$ conductor at the space node.

The reduced Oppenheim network shown in Figure 5-10 represents a network with all conductors defined and given numerical values. The unknowns are the potentials T_{Rad}^4, J_R and J_W. Energy balances are written at each of these points, resulting in the following set of linear equations.

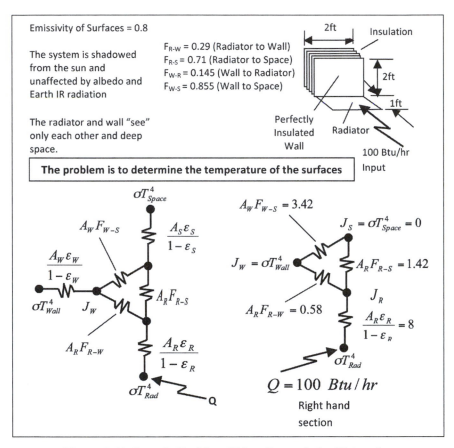

Figure 5-10. Sample Problem Using Oppenheim Networks.

$$100 + 8\left(J_R - \sigma T_{Rad}^4\right) = 0 \tag{5-24}$$

$$8\left(\sigma T_{Rad}^4 - J_R\right) + 0.58(J_W - J_R) + 1.42(0 - J_R) = 0 \tag{5-25}$$

$$0.58(J_R - J_W) + 3.42(0 - J_W) = 0 \tag{5-26}$$

The solution of this set of equations results in:

$$\sigma T_{Rad}^4 = 64.69$$
$$J_R = 52.19$$
$$J_W = \sigma T_{Wall}^4 = 7.568$$

Therefore:

$$T_{Rad} = \left(\frac{64.69}{0.1714 \times 10^{-8}} \right)^{\frac{1}{4}} = 440.8°R$$

GENERALIZED NETWORK ANALYSIS

Radiation networks have been introduced. Heat transfer networks are also helpful in solving conduction and convection problems. We will develop a generalized network approach for solving problems with combined conduction, convection, radiation, fluid flow, and internal heat generation. Internal heat generation includes radiation absorption on a surface and heating by an electrical resistor. An example of a heat transfer network representing conduction and convection is given in Figure 5-11.

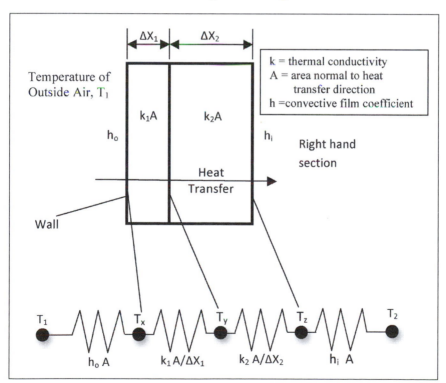

Figure 5-11. Conduction/Convection Thermal Network

In Figure 5-11 a wall is shown made of two materials of different thickness. The bold dots labeled T_1, T_x, etc are nodes. Nodes represent locations in the problem and junctions for the thermal conductors. Nodes T_1 and T_2 are boundary nodes and these nodes are given a temperature as part of the problem statement. There is an outside and inside film coefficient. The quantities shown adjacent to the resistors (h_oA, $k_1A/\Delta X$, etc.) are conductor values, the reciprocal of resistance. Let us assume the outside temperature is 120°F and the inside temperature is 70°F. Also, it will be assumed that:

$h_o = 2$ Btu/(hr-ft2-°F)
$h_i = 1$ Btu/(hr-ft2-°F)
$A = 1$ ft2
$k_1 = .02$ Btu/(hr-ft-°F)
$k_2 = .04$ Btu/(hr-ft-°F)
$\Delta X_1 = 0.5$ ft
$\Delta X_2 = 2$ ft

The conductor values are as follows:

$h_oA = 2$ Btu/(hr-°F)
$h_iA = 1$ Btu/(hr-°F)
$k_1A/\Delta X_1 = .04$ Btu/(hr-°F)
$k_2A/\Delta X_2 = .02$ Btu/(hr-°F)

The problem is steady state, therefore, the heat flowing through each conductor is equal. Because of this we can write the following equations:

$$h_oA\,(T_1 - T_x) + k_1A/\Delta X_1(T_y - T_x) = 0$$
$$k_1A/\Delta X_1(T_x - T_y) + k_2\,A/\Delta X_2(T_z - T_y) = 0$$
$$k_2A/\Delta X_2(T_y - T_z) + h_i\,A(T_2 - T_z) = 0$$

Putting in the numbers, we have:

$2 (120 - T_x) + 0.04(T_y - T_x) = 0$
$0.04(T_x - T_y) + 0.02(T_z - T_y) = 0$
$0.02(T_y - T_z) + 1 (70 - T_z) = 0$

Rearranging the equations:

$-2.04 \, T_x + 0.04 \, T_y = -240$
$0.04 \, T_x - 0.06 \, T_y + 0.02 \, T_z = 0$
$0.02 \, T_y - 1.02 \, T_z = -70$

The set of equations given above can be easily solved since there are three equations and three unknowns. The values for the temperatures are $T_x = 119.67°F$, $T_y = 103.33°F$, and $T_z = 70.65°F$. Another way of solving the equations is by relaxation. For this approach, solve the previous equations for T_x, T_y, and T_z as follows:

$T_x = (0.04 \, T_y + 240)/2.04$
$T_y = (0.04 \, T_x + 0.02 \, T_z)/0.06$
$T_z = (0.02 \, T_y + 70)/1.02$

Guess at values of the temperatures, say, $T_x = 120$, $T_y = 100$, and $T_z = 70$. Use these values to solve the previous equations to obtain $T_x = 119.60$, $T_y = 103.33$, and $T_z = 70.58$. Repeat this process using the newly computed values of the temperatures. In five iterations (repeated computations), these values from the relaxation process will equal the values from solving the simultaneous equations to four place accuracy.

We will now derive an equation for the fluid temperature as a function of the distance along a tube. Consider an energy balance at the fluid element shown below:

An energy balance on the fluid element results in the following equation:

$$M_{dot} \, C_p \, T - M_{dot} \, C_p \, (T + dT) + h \, \pi \, D \, dx \, (T_w - T) = 0$$

Where,

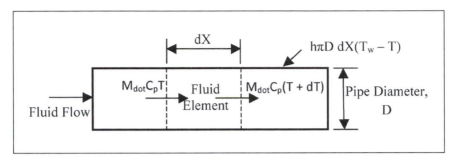

Figure 5-12. Fluid Flow Element

M_{dot} = mass flow rate of the fluid = volume flow rate times fluid density
C_p = specific heat of the fluid
T = fluid temperature
dT = differential temperature change
h = film coefficient
D = tube inside diameter
dx = differential length of tube
$\pi D\, dx$ = surface area of tube at fluid element
T_w = tube wall temperature (a constant)

The energy balance equation reduces to:

$$-M_{dot}\, C_p\, dT = -\, h\, \pi\, D\, dx\, (T_w - T) \tag{5-27}$$

We will approximate the infinitesimal equation, above, with a finite difference equation by letting $dT \approx T_{x+\Delta x} - T_x$ and $dx \approx \Delta x$. Rearranging the terms gives:

$$M_{dot}\, Cp\, (T_x - T_{x+\Delta x}) + h\, \pi\, D\, \Delta x(T_w - T_{x+\Delta x}) = 0 \tag{5-28}$$

Equation (5-27) can be represented by the thermal network shown in Figure 5-13.

The $M_{dot}\, C_p$ conductors shown in Figure 5-13 are one way conductors. This is consistent with Equation (5-28), which shows that node $x + \Delta x$ is influenced only by node T_x and node T_w but

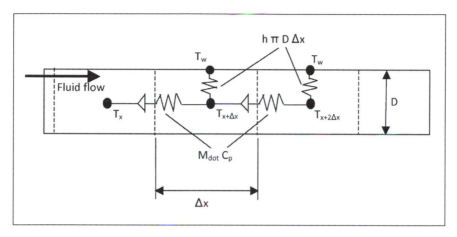

Figure 5-13. Thermal Network For Fluid Flow in a Tube

not by node $T_{x+2\Delta x}$. Note that T_x is an upstream node relative to node $T_{x+\Delta x}$ while node $T_{x+2\Delta x}$ is downstream to node $T_{x+\Delta x}$.

We will now solve a sample problem numerically, and compare the solution with the analytical solution of Equation (5-27). The thermal network which characterizes the problem is given in Figure 5-14, and the quantities that define the problem are given is Table 5-1. A comparison of the numerical and analytical solutions is shown in Table 5-2. The numerical solution is fairly close to the analytical solution and can be made more accurate by increasing the number of nodes.

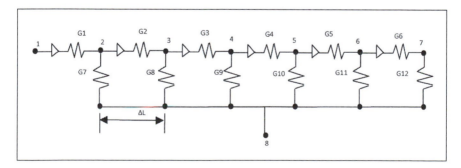

Figure 5-14. Thermal Network For Sample Problem, Steady State Fluid Flow in a Tube

Table 5-1. Material properties, temperatures, and dimensions for the steady state, fluid flow example problem

DENSITY, LBm/CU FT	61.99
VISCOSITY, LBm/(FT-HR)	1.6500
CONDUCTIVITY,BTU/(HR-FT-°F)	0.3640
SPECIFIC HEAT, BTU/(LBm-°F)	0.997
PRANDTL NUMBER	4.519
TUBE DIA, IN	1
TUBE DIA, FT	0.0833333
FLUID VEL., FT/SEC	1
FILM COEF., BTU/HR-FT^2-F	100
TUBE LENGTH, IN	120
TUBE LENGTH, FT	10
DELTA L, IN	20
DELTA L, FT	1.6666667
MASS RATE OF FLOW, LB/SEC	0.338103
REYNOLDS NUMBER	11270.909

Table 5-2. Numerical and analytical solution of the steady state, fluid flow problem

	NUM SOLUTION °F	ANALY SOLUTION °F	% DIFF	DISTANCE FROM INLET, IN
T1	100	100	0	0
T2	103.4708	103.53172	1.72%	20
T3	106.8211	106.93871	1.69%	40
T4	110.0552	110.22538	1.66%	60
T5	113.177	113.39596	1.63%	80
T6	116.1904	116.45458	1.61%	100
T7	119.0993	119.40517	1.58%	120

We will complete the discussion of heat transfer networks by solving the problem illustrated in Figure 5-15. The heat transfer network that models the problem is given in Figure 5-14. Note the problem geometry has been divided into two sections, a left and right section. When problems are solved numerically, as we are doing, the problem geometry can be divided as much as one would desire. The more pieces, or nodes, the more accurate the solution. However, the more pieces the more work required. The items given in Figure 5-16 are defined as follows:

Q = absorbed solar radiation on surface 2 and 9 = 2000 Btu/hr

G1 = h A = film coef * area = 1.0 * 10 = 10 Btu/(hr-°F)

G2 = k A/Δx = thermal cond. * area/thkns = .04 * 10/.25 = 1.6 Btu/(hr-°F)

G3 = h A = film coef * area = 20 * 10 = 200 Btu/(hr-°F)

G4 = h A = film coef * area = 2 * 10 = 20 Btu/(hr-°F)

H = Mdot Cp = mass rate of flow * Specific heat = 200 * 0.24 = 48 Btu/(hr-°F)

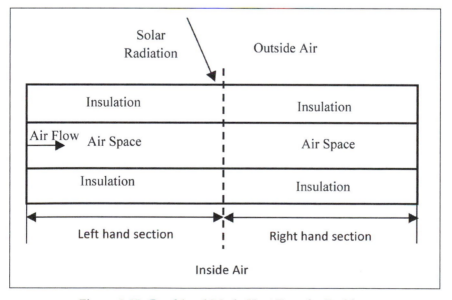

Figure 5-15. Combined Mode Heat Transfer Problem

R1 $= \varepsilon A/(1-\varepsilon) = $ emissivity * area/$(1 - $emissivity$) = .5*10/$
       $(1-.5) = 10$ ft^2
R2 $= $ view factor * area $= 0.6 * 10 = 6$ ft^2
R3 $= $ view factor * area $= 0.05 * 10 = 0.5$ ft^2

The nodes shown in Figure 5-16 are numbered from 1 to 16. The thermal potentials at the nodes are temperatures except when considering an "R" conductor. The thermal potentials relative to R conductors are the Stefan-Boltzmann constant times the absolute temperature to the fourth power. The problem we are considering is steady state. Therefore, we can sum the energy flow at each node point, except the boundary nodes 1, 16 and 17, and set the sum equal to zero. Examples of this are given below:

Node 2
$$Q + G1 * (T1-T2) + G2 * (T3-T2) = 0$$
$$T2 = (Q + G1 * T1 + G2 * T3)/(G1 + G2) \tag{5-29}$$

Node 4
$$G3 * (T3-T4) + G3 * (T5-T4) + H * (T17-T4) = 0$$
$$T4 = (G3 * T3 + G3 * T5 + H * T17)/(G3 + G3 + H) \tag{5-30}$$

Note that node 11 does not affect node 4 since the conductors labeled H are one way fluid conductors.

Node 3:
$$G2 * (T2-T3) + G3 * (T4-T3) + R1 *(\sigma* T7^4-\sigma* T3^4) = 0$$
$$R1* \sigma* T3^4 + (G2 + G3) *T3-G2 * T2-G3 * T4-R1*\sigma * T7^4 = 0$$
$$T3^4 + [(G2 + G3) *T3-G2 * T2-G3 * T4-R1*\sigma * T7^4]/R1 * \sigma = 0 \tag{5-31}$$

Node 7:
$$R1* (\sigma*T3^4-\sigma T7^4) + R2* (\sigma*T8^4-\sigma*T7^4) + R3* (\sigma*T15^4-\sigma *T7^4) = 0$$
$$\sigma T7^4 *(R1 + R2 + R3) = R1* \sigma*T3^4+ R2* \sigma*T8^4 + R3* \sigma*T15^4$$
$$T7^4 = (R1* T3^4+ R2* T8^4 + R3* T15^4)/(R1 + R2 + R3)$$
$$T7 = [(R1* T3^4+ R2* T8^4 + R3* T15^4)/(R1 + R2 + R3)]^{1/4} \tag{5-32}$$

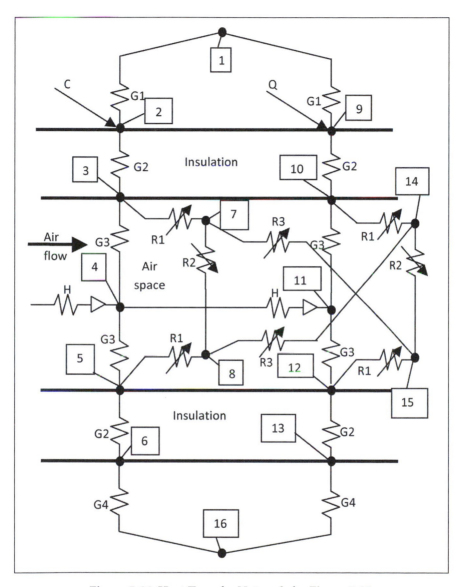

Figure 5-16. Heat Transfer Network for Figure 5-15

To solve the network shown in Figure 5-16, perform an energy balance on each of the non-boundary nodes and derive equations like Equations (5-29), (5-30), (5-31), and (5-32). Initially assume values for all the non-boundary nodes—just make your best guess. Using these guessed values, solve the equations at each node to obtain a new estimate for the node temperature. Repeat this process until the temperature values stop, or very nearly stop, changing. Notice that Equation (5-32) is a quartic equation. In order to solve quartic equations, the reader can use the general solutions on the included CD in the computations folder. It is relatively easy to copy this solution and incorporate it in spreadsheets as needed. This is illustrated in the file entitled *Heat Transfer Network* given in the solutions folder. This file contains a solution of the problem shown in Figure 5-15.

CLOSING REMARKS

In this review of heat transfer, many topics have been omitted. There are several texts available which cover the field of heat transfer extensively. Some of these are general, covering all aspects of heat transfer and others cover a single mode of heat transfer, for example, radiation or conduction. Heat transfer texts are listed in references 5 and 8 through 16.

Chapter 6

Solar Collectors

The purpose of a solar collector is to capture the sun's radiant energy. This energy is then transported to something that will benefit from the energy. The basic configuration of a solar collector includes a surface that collects the sun's heat and transfers this heat to a working fluid, such as water or air. For example, the solar energy may be used to heat domestic hot water, to heat your home or to heat a swimming pool.

There are several types of solar collectors. The most common are flat plate collectors, vacuum tube collectors, and concentrating collectors. An example of a vacuum tube collector is shown in Figure 6-1. The space between the inner and outer tube is a vacuum. The vacuum insulates the inner glass tube (the absorber) from the ambient environment. The surface of the inner tube is coated with a solar absorbing material. This material is usually a so-called selective material. For this type of application, a selective surface material would have a high solar absorptivity, around 0.85, and a low IR emissivity*, around 0.1. Vacuum tube solar collectors are efficient devices meaning that much of the solar radiation hitting the panel is absorbed and "used" by the collector but are relatively expensive. Another type of collector is a concentrating collector, which is shown in Figure 6-2. This type of solar collector is efficient as well and can yield high temperatures to the working fluid. However, concentrating collectors need to be rotated (articulated) in order to keep the solar energy focused on the absorber tube. The most common type of solar collector is the flat plate collector. An example of a flat plate collector is shown in Figure 6-3. The typical size of a flat plate solar collector is about 3 by 8 ft.

---

*Absorptivity is the percentage of the incoming solar energy that is absorbed by the surface, which is receiving the solar energy. Emissivity is a measure of a surface's ability to emit radiation energy.

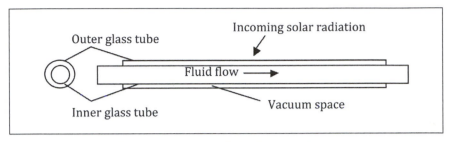

Figure 6-1. Vacuum Tube Solar Collector

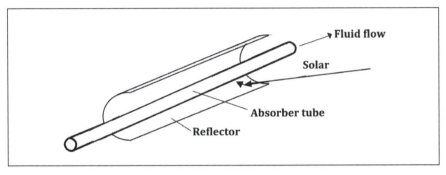

Figure 6-2. Concentrating Solar Collector

The cover is a low iron glass which has a relatively high solar transmissivity. This means that most (about 90%) of the incoming solar energy is transmitted through the glass to the absorber plate. Fluid passages are embedded in the absorber plate. Fluid is pumped through the absorber passages capturing the solar energy and transferring it to its end use in the solar system. The absorber plate is coated with either a selective coating or a black coating. A selective coating absorbs around 80% of the incoming solar radiation but radiates a small amount of energy from the absorber surface, roughly 10%. A black surface absorbs more solar energy than a selective surface, but also radiates more infrared energy than the selective surface. A collector with a black coating on its absorber plate will have a lower stagnation temperature* than a collector

*Stagnation temperature is the temperature on the collector absorber when there is no fluid flow through the collector absorber plate and a maximum amount of solar energy is striking the collector surface.

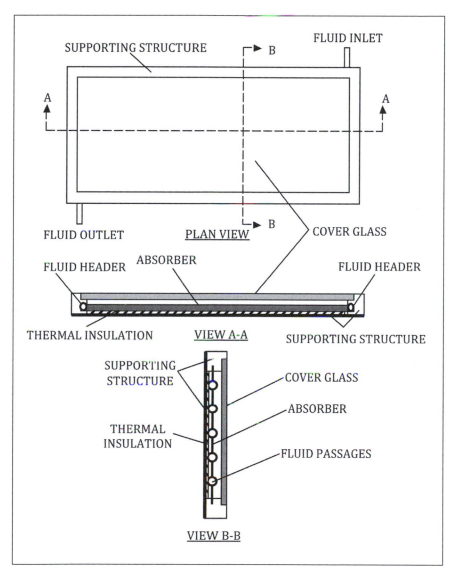

Figure 6-3. Flat Plate Solar Collector

with a selective coating. The stagnation temperature is important since materials used to construct the collector absorber plate must be able to withstand the stagnation temperature. Materials away from the absorber plate must withstand temperatures caused by the stagnation temperature at the absorber plate.

Tubes imbedded in the absorber plate carry a liquid through the collector, picking up the solar heat and transmitting the heat to where the heat is used. There are liquid collectors and air collectors. If freezing is a problem where the collector is installed, the liquid used must have a sufficiently low freezing point. If freezing is not a problem, water is typically used as the working fluid. Air collectors typically do not have imbedded tubes in the absorber plate, but have an air space on the inside of the absorber plate. Often there are fins extending from the inside of the absorber plate and air is pumped through the air space.

We will now discuss liquid flat plate solar collector efficiencies. This is needed to calculate the solar energy absorbed by a flat plate collector. The efficiency defines the performance of the collector to determine the amount of solar energy absorbed by the collector's working fluid. A typical efficiency curve is given in Figure 6-4. In Figure 6-4 the efficiency is the absorbed solar energy divided by the incoming solar energy. T_i is the inlet fluid temperature. T_a is the ambient (outside temperature). I is the incoming solar energy normal to the collector surface (perpendicular to the collector surface). The curve shown in Figure 6-4 can be represented by an equation as follows.

$$\text{Eff} = Q/I = F'\tau\alpha - F'UL^*(T_i - T_a)/I \qquad (6\text{-}1)$$

Where:

Eff	=	the collector efficiency
Q	=	solar energy absorbed by the collector, Btu/(hr-ft^2)
I	=	solar radiation incident on the collector, Btu/(hr-sq ft)
F'τα	=	Efficiency when $(T_i - T_a)/I$ equals zero (y-intercept)
F'UL	=	negative of the slope of the efficiency curve, Btu/(hr-sq ft-F)

T_i = the fluid inlet temperature, F

T_a = the ambient temperature, F

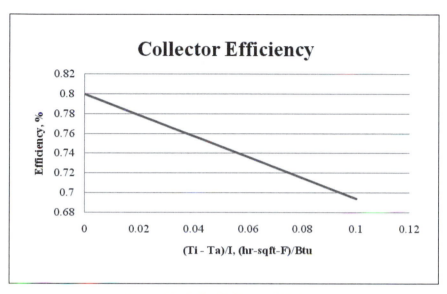

Figure 6-4. Solar Collector Efficiency

Equation (6-1), which is illustrated in Figure 6-4, is not the only efficiency equation for flat plate collectors. Another equation is given by:

$$\text{Efficiency} = [\text{Efficiency at } (T_i - T_a)/I = 0]$$
$$- a_1 (T_i - T_a)/I - a_2 (T_i - T_a)^2/I$$

Where a_1 and a_2 are constants and I, T_i and T_a are as previously defined.

Also, there are collector efficiency curves with collector efficiency plotted versus T_i–T_a and others with collector efficiency plotted versus the average collector fluid temperature.

It is the opinion of the author that Equation (6-1) is the best representation of flat plate solar collector efficiencies. If one considers a given flat plate collector, the variables which most affect the performance of the collector are the fluid inlet temperature, the ambient temperature, the incident solar radiation falling on the collector surface, the wind velocity, and the collector fluid flow rate. The wind velocity and the collector fluid flow rate have a relatively weak effect on the collector efficiency compared to the other variables. Therefore, Equation (6-1) contains the variables that have the strongest effect on collector performance. There are flat plate solar collectors that do not have cover plates. This type of collector is certainly affected by wind and will be discussed later in this section.

The author has developed programs that model the performance of flat plate solar collectors. These models are based on a technical paper by the author given in reference 17. The Microsoft Excel collector programs (one, two or no transparent covers) are given in the *Flat Plate Collector* folder in the Computations folder of the included CD. On each of the collector files, there are the thermal networks used to solve for the collector temperatures. Also, the heat inputs at the collector locations are defined, which account for the absorbed radiation at the various collector parts. Features of the collector programs are given in Figure 6-5.

---

1. All solar & IR band multi-reflections/transmissions accounted for.
2. Internal convection based on experimental data. (See reference 18)
3. Thermophysical property data varied with temperature relative to internal air gap convection and working fluid film coefficient calculations.
4. External convection dependent on wind velocity.
5. Each transparent cover can be assigned its own set of radiation properties (solar transmissivity, reflectivity, absorbtivity, and IR transmissivity, reflectivity, emissivity)
6. Model checked against solar collector test data.

Figure 6-5. Solar Collector Model Features

To test the model against real life, the model was compared to a test of a flat plate collector. Figure 6-6 shows the results of a test on a Thermo Dynamics Ltd., G Series solar collector and the predictions from the solar collector model. The solar collector model was programmed with the physical characteristics and the test conditions for the Thermo Dynamics Ltd., G Series solar collector. The results from the model were remarkably close to the collector test, which is somewhat surprising to the author. Other comparisons of test data and model predictions were reported in Reference 17. For these comparisons, the percent differences between the collector efficiencies measured from tests and the efficiencies derived from the model were 0.3% and 2.0%.

The collector models are useful not only to predict the thermal performance of collectors but also to estimate the stagnation temperature of solar collectors. With the collector model, one not only estimates the maximum temperature on the collector's absorber plate, but also the temperatures throughout the collector. Having these maximum temperatures is important with regard to material selection for the various collector parts.

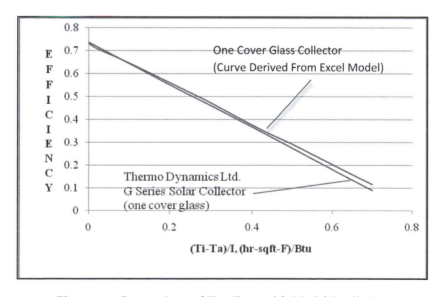

Figure 6-6. Comparison of Test Data with Model Prediction

Figure 6-7 compares the efficiency curves for flat plate collectors with one and two glass cover plates. The extra glass cover absorbs and reflects the incoming solar energy, reducing the efficiency at low values of $(T_i-T_a)/I$. However, at high values of $(T_i-T_a)/I$ the extra glass cover reduces heat losses and yields a higher efficiency than the single cover glass collector.

Efficiency curves for a flat plate solar collector with no cover glass and one inch of back insulation are shown in Figure 6-8. Because there is no cover glass, the wind has a strong influence on the collector performance. For each of the curves shown in Figure 6-8, the y intercept and the slope were determined. These items were plotted versus wind velocity and a quadratic equation fitted to the data. The resulting equation is given below:

$$\text{Eff} = (0.001\ V^2-0.0344\ V + 0.5109) +$$
$$(0.0019\ V^2 -0.0661\ V - 0.8945)\ X$$

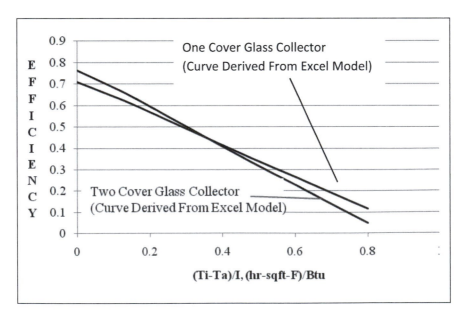

Figure 6-7. Comparison of One and Two Cover Glass Collectors

Where:

V= wind velocity in mph

X= $(T_i - T_a) / I$

T_i= inlet temperature, F

T_a= ambient temperature, F

I= incident solar radiation on collector, Btu/hr-ft^2

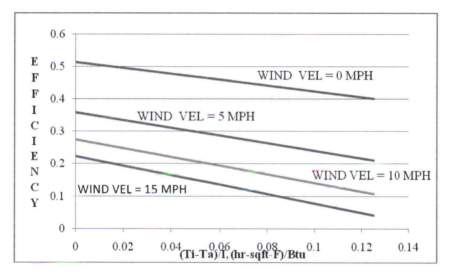

Figure 6-8. No Cover Glass Flat Plate Solar Collector (Data from Excel Model)

Table 6-1. Solar Collector Efficiency Characteristics

Collector Manufacturer	Model Number	F'τα (y-intercept)	F'UL (neg of slope), (hr-ft^2-F)/Btu	Collector Type
Thermo Dynamics Ltd	G series	0.738	0.924	Flat Plate
(Computer Model)	1 cover glass	0.733	0.881	Flat Plate
(Computer Model)	1 cover glass	0.77	0.9	Flat Plate
(Computer Model)	2 cover glass	0.714	0.746	Flat Plate

Efficiency equations for all the collectors shown in the Figures 6-6 through 6-7 are given in Table 6-1.

The last item to cover in this section is the heat transfer at the fins of the absorber plate. A cross section of a typical absorber plate is shown in Figure 6-9.

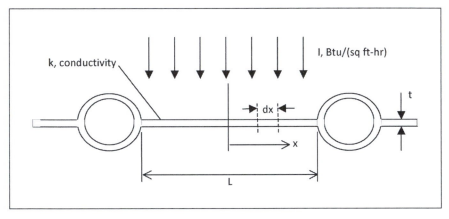

Figure 6-9. Absorber Fin

Performing an energy balance (heat in equals heat out) on the dx element we obtain:

$$I\,dx - k\,t\,dT/dx + k\,t\,[dT/dx + (d^2T/dx^2)dx] = 0 \qquad (6\text{-}2)$$

Equation (6-2) reduces to:

$$d^2T/dx^2 = -I/(k\,t)$$

Solving, we have:

$$T = -I\,x^2/(2\,k\,t) + C_1\,x + C_2$$

Where C_1 and C_2 are constants of integration.

The boundary conditions are:

At x = 0, dT/dx = 0

At x = L/2, T = T_b

Using the boundary conditions to determine C_1 & C_2, we have:

$$T = T_b + I(L^2/4 - x^2)/(2\ k\ t) \tag{6-3}$$

Now the question to ask is, "What is the average temperature along the fin?" This average temperature corresponds to the absorber temperature in the collector Excel models, and is determined by the following equation:

$$T_{avg} = \frac{\int_0^{\frac{L}{2}} T dx}{\frac{L}{2}} \tag{6-4}$$

T_{avg} is the average temperature along the fin.

Substituting Equation (6-3) into Equation (6-4) and integrating we obtain:

$$T_{avg} = T_b + I\ L^2/(12\ k\ t) \tag{6-5}$$

We are looking for the thermal conductor which represents the heat flow in our collector model from the average absorber temperature, T_{avg}, to the base of the fin at the absorber fluid channel, T_b. The heat transfer from the average absorber temperature, T_{avg}, to T_b is given by:

$$Q = (T_{avg} - T_b)\ (k\ t)/\Delta x$$

This heat transfer is equal to the radiation absorbed by the fin, therefore.

$$Q = (T_{avg} - T_b)\ (k\ t)/\Delta x = I\ L/2 \tag{6-6}$$

Substituting Equation (6-5) into Equation (6-6) and solving for Δx we have:

$$\Delta x = L/6$$

In the collector models, this is the value used in the thermal conductor connecting the absorber temperature to the fluid channel wall. Actually, this conductor is combined with the conductor, which represents the area of the fluid channel times the convective film coefficient* for the collector fluid. This conductor connects the absorber to the fluid and is given by:

$$G = \cfrac{1}{\cfrac{1}{hPM} + \cfrac{L}{12ktM}}$$

Where:
- G = thermal conductor connecting the absorber to the fluid channel wall, Btu/(hr-F)
- h = film coefficient, Btu/(hr-sq ft-F)
- P = perimeter of collector tube, ft
- M = total length of tubes, ft
- L = length of collector fins, ft
- k = thermal conductivity of fin material, Btu/(hr-ft-F)
- t = thickness of fin, ft

Note that thermal networks, definitions of conductor values, and heat input values are given in the Excel models for the no glass, one glass and two glass collector models.

---

*The term film coefficient is detailed in Chapter 5's convection section. Film coefficients are solved for in the Computations folder in the CD.

Chapter 7

Absorbed Solar Energy

In this chapter, the information from chapters four (Sunlight on the Earth) and six (Solar Collectors) are combined so that the monthly average solar energy absorbed by the solar collector can be estimated. We start with Equation 4-10 from Chapter 4, which is repeated below:

$$I = I_{DN} [C_N \cos \theta + C/C_N * 0.5 (1 + \cos S)] \qquad (4\text{-}10)$$

(If $\theta > 90°$, set $\cos \theta = 0$)

Where:

I = total solar radiation energy on a surface for an average clear day, Btu/(hr-ft^2)

I_{DN} = direct normal solar radiation (Equation 6-1)

C_N = clearness number

C = indirect coefficient (Table 4-3)

S = tilt angle of the surface, degrees

Next we copy Equation 6-1 from Chapter 6:

$$Eff = Q/I = F'\tau\alpha - F'UL*(T_i - T_a)/I \qquad (6\text{-}1)$$

Where:

Eff = the collector efficiency

Q = solar energy absorbed by the collector, Btu/(hr- ft^2)

I = solar radiation incident on the collector, Btu/(hr- ft^2)

$F'\tau\alpha$ = Efficiency when $(T_i - T_a)/I$ equals zero (y-intercept)

$F'UL$ = negative of the slope of the efficiency curve, Btu/(hr-ft^2-F)

Rewriting Equation 6-1 we obtain:

$$Q = I^* \, F'\tau\alpha - F'UL^*(T_i - T_a), \, Btu/(hr\text{-}ft^2) \tag{7-1}$$

Equation 7-1 is integrated from sunup to sundown to yield the absorbed solar energy, Btu/sq ft, on an average clear day.

$$Q_d = \int_{sundown}^{sunup} (I * F'\tau\alpha - F'UL * (T_i - T_a))d\tau, \, Btu/ft^2\text{-day} \tag{7-2}$$

τ = time

An estimate for the monthly average solar energy absorbed by the solar collector is given by Equation 7-3.

$$Q_m = Q_d * Number \, of \, days \, in \, the \, month * PPSS,$$
$$Btu/ft^2\text{-month} \tag{7-3}$$

PPSS = Percent of Possible Sunshine

Percent of Possible Sunshine is just as the name implies. This term is the actual monthly average sunshine at a given location divided by the monthly average clear sky sunshine at the same location.

Four Microsoft Excel programs have been devised which use Equation 7-3 and provide the following data:

1. I, the total solar radiation incident on a collector surface for an average clear day and for each month, Btu/day

2. I_M, the monthly solar radiation incident on a collector surface (includes the percent of possible sunshine for the month), Btu/month

3. I_Y, the yearly sum for item 2 above, Btu/year

4. Q_A, the total monthly solar radiation absorbed by the collector's working fluid (includes the percent of possible sunshine

for the month and the collector efficiency), Btu/month

5. Q_{AY}, the yearly sum for item 4 above, Btu/year

The information provided in the Excel files can be used in many different ways, including an estimate of the solar energy incident on a surface and the solar energy absorbed on a solar collector. The four programs are contained in the folder entitled, *Incident-Absorbed Solar Radiation*. The programs are:

1. Fixed—The collector is fixed in space

2. Two Axis Tracking—The collector continuously tracks the sun so that the normal to the collector's surface is parallel to the incoming solar radiation

3. One Axis Azimuth Angle Tracking—The collector tilt angle is fixed and the collector's azimuth angle is changed continuously to minimize the angle between the normal to the collector surface and the incoming solar radiation

4. One Axis Tilt Angle Tracking—The collector azimuth angle is fixed and the collector's tilt angle is changed continuously to minimize the angle between the normal to the collector surface and the incoming solar radiation

In order to understand the analyses for modeling the tracking, we need to consider several angles involving the collector surface and the location of the sun. The most important angle is θ, the angle between the normal to the collector's surface and the incoming solar radiation. This angle was defined previously by Equation (4-2) and is repeated below:

$$\cos \theta = \sin \Delta \sin \varphi \cos S - \sin \Delta \cos \varphi \sin S \cos \gamma +$$
$$\cos \Delta \cos \varphi \cos S \cos \omega + \cos \Delta \sin \varphi \sin S \cos \omega \cos \gamma +$$
$$\cos \Delta \sin S \sin \gamma \sin \omega \qquad (4\text{-}2)$$

Where:

θ = the angle between the normal to the collector surface and the solar beam (solar vector), degrees

φ = latitude, degrees

Δ = declination (the angular position of the sun at solar noon with respect to the plane of the equator) declination is determined from an equation or tables, degrees

S = slope of the collector from the horizontal, degrees

γ = azimuth angle (the angle between the direction the collector face is pointing and south, east of south positive & west of south negative), degrees

ω = hour angle (solar noon is zero and each hour equaling 15° of longitude with mornings positive and afternoons negative), degrees. For example, 14:30 = –37.5°.

The solar altitude angle was previously defined by Equation (4-8):

$$\text{Sin } \lambda = \sin \Delta \sin \varphi + \cos \Delta \cos \varphi \cos \omega$$

λ = the solar altitude angle, degrees   (4-8)

Finally, the solar azimuth angle is defined by Equation (7-4):

$$\text{Sin } \eta = \cos \Delta \sin \omega / \cos \lambda \qquad (7\text{-}4)$$

η = solar azimuth angle, degrees

For two axis tracking, set the tilt angle equal to 90— λ (the solar altitude angle) and set the azimuth angle (γ) equal to the solar azimuth angle (η). This causes θ to equal 0 and cos (0) =1.

For one axis azimuth angle tracking rearrange Equation (4-2) as follows:

$$\cos \theta = \sin \Delta \sin \varphi \cos S + \cos \Delta \cos \varphi \cos S \cos \omega +$$
$$\cos \gamma(\cos \Delta \sin \varphi \sin S \cos \omega - \sin \Delta \cos \varphi \sin S) +$$
$$\sin \gamma(\cos \Delta \sin S \sin \omega) \qquad (7\text{-}5)$$

Let:

$C = \sin \Delta \sin \varphi \cos S + \cos \Delta \cos \varphi \cos S \cos \omega$

$C1 = \cos \Delta \sin \varphi \sin S \cos \omega - \sin \Delta \cos \varphi \sin S$

$C2 = \cos \Delta \sin S \sin \omega$

Substituting C, C1, and C2 into Equation (7-5) we have:

$$\cos \theta = C + C1 \cos \gamma + C2 \sin \gamma \qquad (7\text{-}6)$$

In order to minimize θ, differentiate $\cos \theta$ with respect to γ and set the result equal to zero.

$$\frac{d \cos \theta}{d\gamma} = - C1 \sin \gamma + C2 \cos \gamma = 0$$

$$\frac{\sin \gamma}{\cos \gamma} = \frac{C2}{C1}$$

$$\tan \gamma = \frac{C2}{C1}$$

$$\gamma = \tan^{-1}\left(\frac{C2}{C1}\right) \qquad (7\text{-}7)$$

Equation (7-7) defines the azimuth angle at each time step which minimizes θ, the angle between the normal to the collector surface and the incoming solar radiation.

For one axis tilt angle tracking, rearrange Equation (4-2) as follows:

$$\cos \theta = \cos S (\sin \Delta \sin \varphi + \cos \Delta \cos \varphi \cos \omega) +$$
$$\sin S (\cos \gamma \cos \Delta \sin \varphi \cos \omega - \sin \Delta \cos \varphi \cos \gamma +$$
$$\sin \gamma \cos \Delta \sin \omega) \qquad (7\text{-}8)$$

Let:

$D1 = \sin \Delta \sin \varphi + \cos \Delta \cos \varphi \cos \omega$

$D2 = \cos \gamma \cos \Delta \sin \varphi \cos \omega - \sin \Delta \cos \varphi \cos \gamma +$
$\quad \sin \gamma \cos \Delta \sin \omega$

Substituting D1, and D2 into Equation (7-8) we have:

$$\cos \theta = D1 \cos S + D2 \sin S \qquad (7\text{-}9)$$

In order to minimize θ, differentiate $\cos \theta$ with respect to S and set the result equal to zero.

$$\frac{d \cos \theta}{dS} = -D1 \sin S + D2 \cos S = 0$$

$$\frac{\sin S}{\cos S} = \frac{D2}{D1}$$

$$\tan S = \frac{D2}{D1}$$

$$S = \tan^{-1}\left(\frac{D2}{D1}\right) \qquad (7\text{-}10)$$

Equation (7-10) defines the tilt angle at each time step which minimizes θ, the angle between the normal to the collector surface and the incoming solar radiation.

It is worth noting that when using the *Incident — Absorbed Solar Radiation* programs if 1 is input for the collector efficiency at the y intercept, F'$\tau\sigma$, and zero is input for the slope of the collector efficiency curve, F'UL, the absorbed solar energy is equal to the incident solar energy. If it is desired to have the absorbed solar energy a fixed fraction of the incident solar radiation, input the fraction into F'$\tau\alpha$ and input zero for F'UL.

This chapter will be concluded with a sample problem. Assume a 100 square foot PV array. This array is located in Fort Worth, Texas and has an average efficiency of 20%. Compare the yearly electrical energy produced by the array if the array is fixed and if the array is one axis tilt angle tracking.

In the *Incident — Absorbed Solar Radiation* folder, open the file 'Fixed'. Next, go to the Weather Data-F file. Fort Worth is not listed so determine the listing which is the closest to Fort Worth. Dallas,

Texas, is about 35 miles from Fort Worth so the Dallas weather data will be used. Also, the latitude of Dallas is close to the latitude of Fort Worth. Highlight the Dallas weather data that are in the region defined by cells A795 and M799. Copy the Dallas weather data and go to the *Fixed* file. Highlight A14 and paste. The data input cells for the *Fixed* program are cells 21, 24, 27 thru 31 all in the B column. The input values are listed in Table 7-1.

Table 7-1. Fixed program input values

Cell	ITEM	VALUE
B21	TILT ANGLE	VARY TO FIND BEST VALUE
B24	AZIMUTH ANGLE	ZERO
B27	F'τα	0.20
B28	F"UL	ZERO
B29	COLLECTOR AREA, sq ft	100
B30	NO. OF COLLECTORS	1
B31	INLET FLUID TEMP	DOES NOT APPLY

Vary the value of the tilt angle starting with the latitude. A value of 26 degrees was found to maximize the power generated at 9.69E6 Btu/year. This is equal to 2,839 kW-hr. The conversion from Btu to kW-hr is 2.93E-4 kW-hr/Btu.

Go to the *Incident-Absorbed Solar Radiation* folder and open the *One Axis Tilt Angle Tracking* file. Next go to the *Weather-F* file. Highlight the Dallas weather data, which are in the region defined by cells A795 and M799. Copy the Dallas weather data and go to the *One Axis Tilt Angle Tracking* file. Highlight A14 and paste. The data input cells for the Fixed program are cells 21, 24 thru 28 all in the B column. The input values are listed in Table 7-2.

The power output for the solar arrays with one axis tilt angle tracking is 3,054 kW-hr. This represents a 7.1% increase in output compared to the fixed array.

Table 7-2. One axis tilt angle tracking program input values

Cell	ITEM	VALUE
B21	AZIMUTH ANGLE	ZERO
B24	F'τα	0.20
B25	F"UL	ZERO
B26	COLLECTOR AREA, sq ft	100
B27	NO. OF COLLECTORS	1
B28	INLET FLUID TEMP	DOES NOT APPLY

Chapter 8

Solar Domestic Hot Water Systems

In this chapter, a Microsoft Excel program is introduced which predicts the savings and payback period for domestic hot water solar systems. This program is in the Microsoft Excel file, *Domestic Hot Water*. A diagram of a hot water solar system is shown in Figure 8-1.

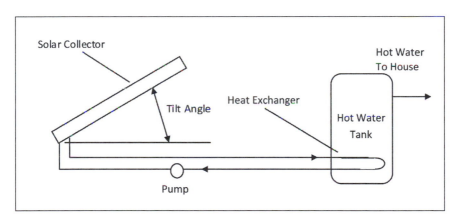

Figure 8-1. Domestic Hot Water Solar System

There are several variations of solar hot water systems. The system shown in Figure 8-1 is often expanded with an additional hot water tank. For cold weather locations, either non-toxic anti-freezes (commonly propylene glycol) are used for the collectors' working fluid, or the systems are set up to drain the water from the collector when the fluid temperature approaches freezing. This latter approach is called a drain back system.

Monthly weather data used in the program include the average daytime temperature, the percent of possible sunshine, and the atmospheric clearness number (from the ASHRAE handbook) of the location being analyzed. The average daytime temperature is assumed to be the average of the daily high temperature and the daily average temperature. There are over 150 US cities in the weather database. The weather data are given in the Microsoft Excel file *Weather Data-F*.

The remainder of this chapter discusses an overall approach for estimating the energy cost savings realized from a solar domestic hot water system; however, it does not include all the details of the calculations. Before beginning the cost savings analysis, the efficiency of the collector(s) must be determined. Solar collector dealers should know the values of F'τα and F'UL. In addition, representative values are provided in Chapter 7, Solar Collectors.

How do you estimate the savings due to a solar domestic hot water system? The material given here does not go through all the details of the calculations but it does describe the overall approach. The cost savings analysis is based on computing the solar energy absorbed by a collector on a clear day at the middle of each month. The clear day absorbed solar energy is adjusted for cloudy days using weather data known as Percent of Possible Sunshine. There is a final adjustment to the monthly average absorbed energy referred to as System Loss Factor (SLF). This factor accounts for energy absorbed during periods when the sun is shining but the hot water is at the maximum temperature allowable so this absorbed solar energy cannot be used, for heat losses from the system piping and tanks, and other miscellaneous losses. The solar energy used for hot water heating equals the clear day absorbed solar energy times the percent of possible sunshine times (1 – System Loss Factor). This mid-month value for the solar energy available for hot water heating is assumed to be representative for the entire month. The System Loss Factor is assumed to be 0.18. This value is based on data from yearly tests on hot water solar systems in Hadley, Massachusetts, and in Madison, Wisconsin. See reference 19 for details.

Begin with one solar collector for the system in the cost savings analysis. Assuming that one collector provides more than enough energy to meet the monthly hot water needs, the savings equal the conventional cost of the hot water usage. In realistic conditions, there are cloudy and partially cloudy days when the solar system can provide only a portion or perhaps none of the hot water requirements. Based on these weather variables, it is assumed that the maximum monthly solar energy absorbed by the solar hot water system equals the monthly hot water requirement times the percent of possible sunshine. Starting the analysis with one collector may not meet the hot water requirements. In this case, note the payback period with one collector and then add a collector and compare the payback period with two collectors. Continue to add collectors until the payback period starts to increase. Each collector added increases the cost of the system, thereby affecting the payback period. The goal is to achieve a balance between the supply and demand of hot water. Another variable in the analysis is the tilt angle of the collector. Begin with an angle equal to the latitude of the city used in the analysis and vary this number to find the tilt angle that maximizes savings.

A domestic solar hot water system sample problem follows. Assume the city is Denver, Colorado; the household is made up of four people, the hot water temperature is 150°F, and the inlet water temperature is 50°F. The current energy source is electricity at a price of 0.1 $/(kW-hr). The basic price of the solar hot water system is $2500 which includes one 24 ft² solar collector. Additional collectors cost $1000 each. The collector F'τα =0. 71 and the F'UL = 0.78 Btu/(hr-sq ft-F).

Open the domestic solar hot water file, *Domestic Hot Water*. Next open the file *Weather Data-F* and find the data for Denver, Colorado. Highlight the Denver weather data that is in the region defined by cells A111 and M115. Copy the Denver weather data and go to the *Domestic Hot Water* file. Highlight A9 and paste.

Enter the data given in the problem statement above into the bordered cells in the *Domestic Hot Water* file. This takes it down through cell B47. Enter zero in cell B49, the azimuth angle and 40 in cell B48, the tilt angle. At this point, assume one collector and enter this in cell B50. Now vary the tilt angle to find the maximum savings. The maximum savings of $242/yr occur at a tilt angle of 34 degrees.

With one collector and a system cost of $2500, the payback is 10.3 years. Change the number of collectors to two and the system cost to $3500. This change yields a pay back of 7.4 years. At a system cost of $4500 with three collectors, the pay back is 8.3 years, so the optimum is the two-collector system.

It is important to understand that the analysis presented here yields approximate values for the cost savings and the payback period. The analysis does take into account weather data, the collector performance, attenuation of the solar energy coming through the atmosphere, the angle between the surface of the collector and the solar beam, and the clearness of the atmosphere. However, the calculations are not exact and, of course, the historic weather data may not represent the weather in the future. Also, the System Loss Factor (discussed above) is inexact. Therefore, use the information given here as a guide and realize the results are certainly not exact.

Chapter 9

Solar Space Heating

In Chapter 7 techniques were developed for estimating the solar energy absorbed by solar collectors. Monthly average daytime temperatures and the percent of possible sunshine values are presented in the file *Weather Data-F* for over 150 US cities. The file *Weather Data-TT* references the same US cities; however, instead of average daytime temperatures, it lists average daily temperatures. Average daytime temperatures are used when analyzing solar collection while the sun is shining. Average daily temperatures are used when analyzing the average heat loss over a 24-hour period, daytime and nighttime.

The following sample problem illustrates how to analyze a solar space heating system. To simplify, consider a one-story, 8-ft-high home located in Cheyenne, Wyoming. The home is rectangular, 36' by 48' with a flat roof. The construction is 2" x 6" frame (on 16" centers) with ¾" exterior plywood covered with ¾" cedar siding. The inside of the exterior walls are covered with ½" dry wall. The exterior walls are filled with 5½" glass wool between the 16" studs. The home is built on a concrete slab. The flat roof is constructed of 2" x 12" rafters with ¾" plywood on the outside and ½" dry wall on the inside. Glass wool (11½" thick) is installed between the rafters. The total area of the double glazed windows is 336 sq ft. The total area of the 2" thick exterior wood doors is 65 sq ft.

To solve this problem, the overall heat transfer coefficients for all elements of the exterior of the home are needed. Overall heat transfer coefficients are defined as follows:

$$Q = UA\Delta T \tag{9-1}$$

Where,

$Q =$ heat transfer, Btu/hr

$U =$ overall heat transfer coefficient, Btu/(hr- sq ft- F)

$A =$ area, sq ft

$\Delta T =$ temperature difference (inside to outside), F

The external wall construction detail, along with a thermal network of the wall, is shown in Figure 9-1.

The overall heat transfer coefficient for the external wall is computed as follows:

$$UA_{wall} = \cfrac{1}{\cfrac{1}{h_o A} + \cfrac{\Delta X_c}{k_c A} + \cfrac{\Delta X_p}{k_p A} + \cfrac{1}{\cfrac{k_{gw} A_{gw}}{\Delta X_{gw}} + \cfrac{k_{ps} A_{ps}}{\Delta X_{ps}}} + \cfrac{\Delta X_{dw}}{k_{dw} A} + \cfrac{1}{h_i A}}$$

(9-2)

Where:

$(UA)_{wall}$ = Overall film coefficient times the wall area, Btu/(hr-°F)

h_o = outside film coefficient, Btu/(hr-ft2-°F) assume $h_o = \infty$ (very large)

A = area of wall (this problem worked on a one ft2 basis) = 1 ft2

ΔX_c = thickness of the cedar siding =(3/4)/12 ft

k_c = thermal conductivity of cedar = 0.0617 Btu/(hr-ft-°F)

ΔX_p = thickness of plywood = (3/4)/12 ft

k_p = thermal conductivity of plywood=0.0617 Btu/(hr-ft-°F)

k_{gw} = thermal conductivity of glass wool = 0.0179 Btu/(hr-ft-°F)

A_{gw} = area of glass wool =(9" x 14 ½")/144 ft2

ΔX_{gw} = thickness of glass wool = 5.5"/12 ft

k_{ps} = thermal conductivity of pine stud = 0.065 Btu/(hr-ft-°F)

A_{ps} = area of pine stud = (9" x 1.5")/144 ft2

ΔX_{ps} = thickness of pine stud = 5.5"/12 ft

k_{dw} = thermal conductivity of dry wall = 0.28 Btu/(hr-ft-°F)

ΔX_{dw} = thickness of dry wall = 0.5/12 ft

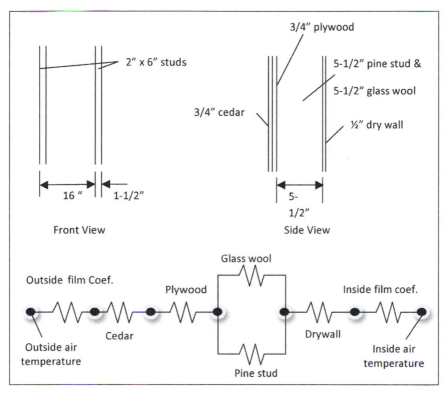

Figure 9-1. Exterior Wall Construction Detail and Thermal Network

h_i = inside film coefficient = 0.8 Btu/(hr-ft2-°F)
$(U)_{wall}$ = 0.0417 Btu/(hr-ft2-°F)

The following table defines all the overall heat transfer coefficients needed to compute the heat loss from the Cheyenne home in the sample problem.

The "UA" for the slab in Table 9-1 was determined from Reference 20. This reference contains a significant amount of information on heat loss calculations and is highly recommended.

The overall heat transfer coefficient for a residence can be estimated from the monthly energy bill for the residence. The amount of heating (in Btu's) for common fuels is given in Table 9-2.

A monthly bill for the heating fuel can be used to estimate the overall heat transfer coefficient for the home. If the heating fuel is

Table 9-1. Overall Heat Transfer Coefficients and Heat Transfer Areas for Sample Problem

Heat Transfer Path	U Btu/(hr-sq ft-F)	Area(A) sq ft	UA, Btu/(hr-F)
Exterior Walls	0.0417	943	39.3
Windows	0.5	336	168
Doors	0.45	65	29.3
Roof	0.02	1728	34.6
Slab	---	---	252
Total Btu/(hr-F)			**523.2**

Table 9-2. Heating Values for Common Fuels

Fuel	**Units**	**BTU's**
Natural Gas	1000 ft^3	919,963
Natural gas	Therms	80,000
Fuel oil	Gallons	82,169
Propane	Gallons	85,543
Electricity	Kw-hr	3413

electricity and a clothes dryer vented to the outside is used, multiply the monthly fuel usage by 0.942. (This value is from DOE). If substantial outside lighting is used, estimate the monthly kW-hrs used by the outside lights and subtract this from the monthly overall usage. Now take the adjusted kW-hrs and multiply by 3413 (from Table 9-2). Finally, use the following equation to estimate the monthly overall heat transfer coefficient times the area:

$$Q = UA(T_o{-}T_i)*(24*\text{number of days in the current month})$$
$$UA = Q/(T_o{-}T_i)/(24*\text{number of days in the current month})$$

$$(9\text{-}3)$$

Where:

Q	=	monthly heating, Btu/mouth
T_o	=	monthly average outside temperature, °F*
T_i	=	average inside temperature, °F
UA	=	overall heat transfer coefficient, Btu/(hr-F)

Perform this calculation for each month that the average outside temperature is less than the inside temperature. Average the UA's to establish an average UA for the home.

If the heating fuel is other than electricity, use the same approach as outlined above except add in the Btu's from the primary heating fuel (Natural gas, fuel oil, or propane) to the electrical Btu's. Note that even if the electrical usage is primarily for lighting, refrigeration, etc., this energy ultimately ends up as heat in the home.

The residence UA (given in Table 9-1) and the program given in the file Solar Space Heating will be used to finish the sample problem. Open the files *Weather Data-F*, *Weather Data-TT*, and *Solar Space Heating*. Copy the data for Cheyenne, Wyoming from the weather data files and paste into the *Solar Space Heating* file. Assume the heating fuel is electricity at a cost of $0.10/kw-hr, the collectors have an area of 24 sq ft with a $F'_{\tau\alpha} = 0.71$ and $F'_{UL} = 0.78$. Starting with one collector, vary the tile angle until the saving ($/year) becomes a maximum. This tilt angle is 40 degrees. Next, increase the number of collectors until the minimum payback period is found. Assume a base cost of $2500 for the system plus $600 for each collector (i.e., a system with one collector costs $3100). The payback period is 6.2 years and the number of collectors is seven.

---

*From the file WEATHER DATA-TT

Chapter 10

Solar Electric Systems

Up to this point, the solar systems discussed in this book have concentrated on systems that capture the sun's radiant energy for the sole purpose of transferring heat to a working fluid. This chapter introduces solar systems that produce electricity—solar power towers and solar photovoltaic systems. A solar power tower concentrates the sun's energy to heat a working fluid that produces steam to generate electricity via a steam turbine. Solar photovoltaic systems convert sunlight directly into electricity.

SOLAR POWER TOWERS

Figure 10-1 shows a molten salt solar power tower system. The solar energy reflected off the heliostats enters the solar receiver and is absorbed by molten salt flowing through the receiver. Heliostats are large mirrors mounted on a two axis tracking system. The mirrors reflect incident solar energy to the receiver. The heated molten salt at the receiver flows to a tank and on to a steam boiler. From the boiler, the molten salt flows to a cool molten salt tank and back to the receiver to complete the molten salt loop. At the boiler, the molten salt supplies the energy to produce steam. The steam moves to a turbine and then to a condenser. The water out of the condenser flows to the boiler. The steam turbine drives an electric generator. The electrical energy from the generator is the output of the system. The boiler, turbine, generator, and condenser arrangement in the solar power tower system is very similar to a conventional power plant except the energy for the boiler is supplied by hot molten salt instead of coal, fuel oil, or natural gas.

Solar power tower systems have been built which use water, liquid sodium, or air as the heat absorbing media at the receiver. The advantage molten salt provides over the other fluids is the ability of the molten salt to store thermal energy. This allows the molten salt solar power tower system to operate and produce electricity when the sun is not shining.

There are many steps between solar energy striking the outer fringe of the Earth's atmosphere to the electrical power from the generator. The thermal radiation, which arrives at the Earth's outer atmosphere is about 430 Btu/(hr-ft^2) and is reduced, on

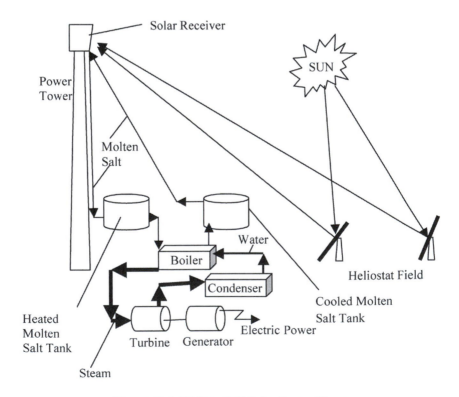

Figure 10-1. Molten Salt Solar Power Tower

a clear day, to about 350 Btu/(hr-ft^2) after passing through the Earth's atmosphere. At the heliostats, there is a reduction in solar energy due to the transmission losses through the mirror cover material, the reflection loss off the mirror, and the reflection angle at the heliostat. The reflection angle and the incident angle are each one half of the angle between the incoming solar beam and the line from the heliostat to the receiver. Figure 10-2 illustrates the heliostat/receiver geometry and this is used to define the line between the heliostat and the receiver. Imagine, at the heliostat, a flat plate collector pointing directly at the receiver and find the azimuth angle and tilt angle of the collector. For the tilt angle, use

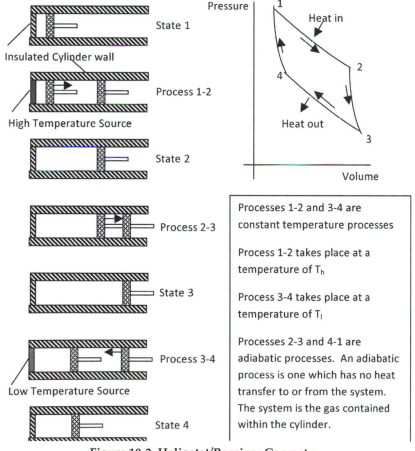

State 1

Insulated Cylinder wall

Process 1-2

High Temperature Source

State 2

Process 2-3

State 3

Process 3-4

Low Temperature Source

State 4

Pressure

Heat in

Heat out

Volume

Processes 1-2 and 3-4 are constant temperature processes

Process 1-2 takes place at a temperature of T_h

Process 3-4 takes place at a temperature of T_l

Processes 2-3 and 4-1 are adiabatic processes. An adiabatic process is one which has no heat transfer to or from the system. The system is the gas contained within the cylinder.

Figure 10-2. Heliostat/Receiver Geometry

positive values for X, Y, and Z. For the azimuth angle, X is positive if the tower is south of the heliostat, and Y is positive if the tower is east of the heliostat. See the quadrant definition in Figure 10-2. With the tower in the 1st and 2nd quadrants, the azimuth angle is negative, and in the 3rd and 4th quadrants, the azimuth angle is positive.

Azimuth angle = γ = $\tan^{-1}(Y/X)$

Tilt angle = S = $90° - \tan^{-1}(Z/K)$ where $K = (X^2 + Y^2)^{1/2}$

With these angles, we can use the previously developed technique to find the angle between the incoming solar beam and the normal to the "collector." This angle will be equal to the sum of the reflection angle and the incident angle at the heliostat's mirror.

To illustrate the method for computing the reflection and incident angles, consider the following sample problem. A heliostat is located 300 ft north of the power tower and 100 ft east of the north/south line through the tower. Referring to Figure 10-2, X = 300 and Y = –100. The height of the center of the receiver aperture above ground level is 200 ft, therefore Z = 200. Find the incident and reflection angle at the heliostat mirror for 11:45 AM on July 23. Assume the system is located in Albuquerque, New Mexico.

First, find the tilt and azimuth angles for the heliostat/receiver line (i.e., the line that points the heliostat's normal directly at the tower) as shown in Figure 10-2:

Azimuth angle = γ = $\tan^{-1}(-100/300)$ = $-18.43°$

Since the tower is in the 1st quadrant the azimuth angle is negative.

$K = (300^2 + 100^2)^{1/2} = 316.2$ ft

Tilt angle = S = $90° - \tan^{-1}(200/316.2)$ = $57.69°$

Next, use these values to find the angle between the solar vector and the heliostat normal. Go to the file Weather Data-F and find the latitude of Albuquerque, New Mexico, which is 35.05°. Next, open the file Sun Location and input the tilt and azimuth angles calculated above, day of the year (204), latitude, and time of day. The angle between the heliostat/receiver line and the solar beam is 45.3°. Therefore, the incident and reflection angles are 22.65° each. The loss due to the reflection is:

$$\text{Loss} = 1 - \cos(22.65°) * \cos(22.65°) = 0.148 = 14.8\%$$

One would not think that there is a measurable attenuation of the solar beams from the heliostats to the receiver; however, there is a small loss. A rough estimate of this loss is made by taking the mass of the atmosphere and then calculating the height of the atmosphere, assuming the density of the atmosphere is constant and equal to the sea level density. This height is 28,937 ft. We know that the solar intensity at the fringe of the Earth's atmosphere is approximately 430 Btu/(hr-ft^2). Also, on a clear day the solar intensity on the Earth's surface is about 350 Btu/(hr-ft^2). Therefore, the loss per ft is $(430–350)/28937 = 0.002764$ Btu/(hr-ft^2)/ft. If the distance from the heliostat to the receiver is 600 ft, the loss in solar intensity would be 1.659 Btu/(hr-ft^2) and assuming the solar intensity at the heliostat is 350 Btu/(hr-ft^2) the percentage loss to the receiver is 0.47%.

Heat loss from a receiver is made up of reflected radiation, emitted radiation, convection, and conduction. One type of receiver is a cavity receiver shown in Figure 10-3. This receiver was tested in the early 1980's at Sandia's National Solar Thermal Test Facility in Albuquerque, New Mexico. Part of the test program was dedicated to determining the convection heat loss from the receiver and was found to be 1.43 percent of the maximum heat input of 5 MW. See Reference 21. Reflection and emission radiation percent losses at the receiver are 1.72 and 2.07, respectively. The analyses used to estimate these losses are shown on the file, receiver F.

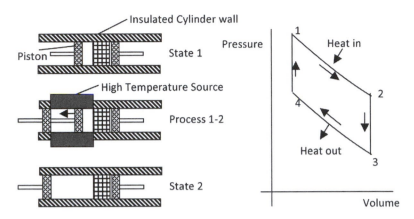

Figure 10-3. Cavity Receiver

Table 10-1 shows the losses from the solar energy at the heliostats to the output of the generator. These values are approximate and are shown to give a general idea of the losses involved in a solar power tower system.

The author was a member of a team that designed and then tested the 5 MW receiver shown in Figure 10-3. During the operation of the system, the molten salt flow rate was varied to cause the outlet molten salt temperature to equal 1050°F. For clear day operation, the changes in the molten salt flow rate to maintain the 1050°F outlet temperature were small. During cloudy day operation, molten salt flow rate changes could be quite large. If the flow rate is adjusted too low, the outlet temperature will rise

Table 10-1. Solar Power Tower Energy Losses

SOLAR RADIATION AT HELIOSTATS	100.0
SOLAR RADIATION AFTER REFLECTION AT HELIOSTAT	90.0
SOLAR RADIATION AFTER "COS" LOSS AT HELIOSTAT	76.5
SOLAR RADIATION ENTERING THE RECEIVER	76.1
ENERGY ABSORBED BY MOLTEN SALT	72.1
ENERGY AT BOILER	71.0
GENERATOR OUTPUT	28.4

above the set point of 1050°F. Conversely, if the flow rate is too high, the outlet temperature will drop below the 1050°F set point. If for some reason the molten salt flow rate is low and the solar from the heliostats is high, the outlet temperature will exceed the 1050°F set point. In the extreme, the solar energy from the heliostats could melt the steel receiver tubes. If there is no heat input from the heliostats, the molten salt is drained from the receiver tubes and stored in a heated tank. It is possible to freeze the salt in the receiver tubes if the draining is not carried out in a timely manner. If the tubes have to be thawed, it is likely they will be broken in the process. It was a relief to finish our testing without melting the receiver tubes or freezing the salt in the tubes!

SOLAR PHOTOVOLTAIC SYSTEMS

Solar photovoltaic (PV) systems use solar cells to convert sunlight directly into electrical energy. Solar cells produce direct current and usually the direct current is inverted to alternating current. The most common solar cell material is silicon. PV systems vary greatly in size and complexity. Small systems with fixed solar panels (solar cell arrays) of one or two square feet are commonly seen along highways powering signs, telephones, etc. Almost all spacecraft have solar arrays that provide the spacecraft with electrical power. At Denver International Airport, there is a PV system covering 7.5 acres that produces approximately 3.5 million kW-hr of electricity annually. This system is a one axis tilt angle tracking system. PV systems have an application in residential electrical power. The usual arrangement is to mount the solar arrays on the roof and invert the direct current to alternating current. If the PV system produces more electrical power than the residence requires, the excess is returned to the power grid and the home owner is credited for this power. In Colorado, the state requires that the credit is the same as the original cost of the power.

A sample problem follows to illustrate how material presented on the CD can be used to investigate PV systems.

Assume a PV system is to be built in Phoenix, Arizona. The power output of the solar arrays is to be 5000 MW-hr/yr. The array overall efficiency is 12%. What is the size of the array if 1) one axis tilt angle tracking is used and 2) fixed arrays are used?

Open the Microsoft Excel file *One Axis Tilt Angle Tracking* from the *Incident –Absorbed Solar Radiation* folder. Next open the *Weather Data-F* file. In this file find Phoenix, Arizona, at line 39 and highlight the area from cells A39 to M43. Copy the highlighted data and paste it into cell A14 in the *One Axis Tilt Angle Tracking* file. In this file enter 0 in cell B21 (the azimuth angle), 0.12 in cell B24 (F'τα-in this case is the system efficiency), 0 in cell B25 (F'UL), 435600 (10 acres) in cell B26 (array area in sq ft), and 1 in B27 (number of collectors). It does not matter much what is used in cell B28 (inlet temperature), so use 45 F. Now vary the collector area until the output is 5.0 MW-hr/yr. The area is 190,900 sq ft.

Repeat the calculations using the *Fixed* file from the Incident –*Absorbed Solar Radiation* folder. For this calculation use an area of 210,000 sq ft and vary the tilt angle to establish the optimum. The optimum tilt angle is 27 degrees. With this tilt angle, vary the area until the output is 5.0 MW-hr/yr. The area required for a fixed array system is 205,600 sq ft. This represents a 7.8% increase over the tracking system.

Chapter 11

Stirling Engine Solar Power Systems

Figure 11-1 shows a Stirling engine solar power system. A solar system of this type uses a parabolic mirror, which tracks the sun to reflect sunlight on a Stirling engine. The engine drives an electric generator, and heat rejected by the engine is delivered to the atmosphere

Before the details of the Stirling engine are discussed, it will be helpful to inspect the Carnot engine and the thermodynamics associated with a Carnot engine. This engine was proposed by Nicholas Leonard Sadi Carnot (1796-1832) in his technical paper *Reflections on the Motive Power of Heat*. The information in this paper is extremely important to the science of Thermodynamics, and serves as a foundation for the Second Law of Thermodynamics. Details of the Carnot engine are shown in Figure 11-2. Note that this type of engine is an external combustion engine as opposed to the ordinary internal combustion engine used in automobiles. Like an automobile gasoline engine, the Carnot engine has cylinders and pistons. However, the gas in the cylinder of a Carnot engine is permanently captured within the cylinder. There are no inlet or outlet valves in a Carnot engine. A Carnot engine has the maximum efficiency possible for a heat engine operating between two constant temperature heat sources.

A heat engine is a device that uses heat as an energy source and converts part of the heat into work. Work is defined as force times distance, and power is work per unit time. Gasoline engines, diesel engines, and steam power plants are examples of heat engines. An important question is, "what is the maximum efficiency of heat engines?" Efficiency is defined as the work output of an engine divided by the heat input to the engine. The Carnot engine

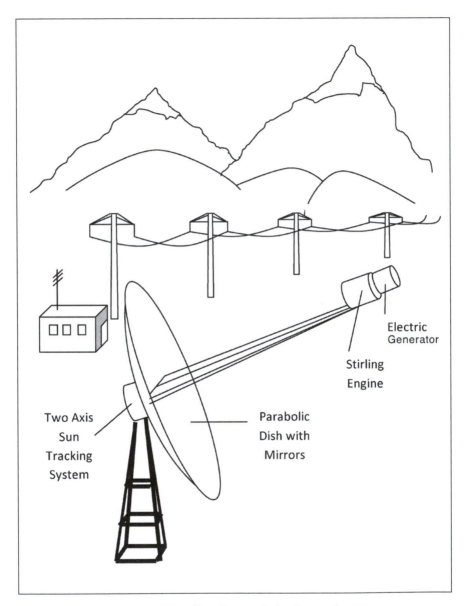

Figure 11-1. Sterling Energy Solar Power System

is a heat engine that receives heat at a constant high temperature and ejects waste heat at a constant low temperature. It turns out that the efficiency of a Carnot engine is:

$$\eta = 1 - T_{low}/T_{high}{}^* \tag{11-1}$$

Where

T_{low} = temperature of the low temperature source in °R†
T_{high} = temperature of the high temperature source in °R

   The processes that take place in a Carnot engine are reversible processes. A reversible process is frictionless and actually unattainable; however, using the concept of reversible processes allows the development of important laws and equations. Also, some processes are very close to being reversible. For a comparison of actual and theoretical thermodynamic processes, see the appendix, "Air Gun-1."

   It can be shown that the efficiency of a Carnot engine is the maximum possible for a heat engine operating between a high and a low temperature source. An example of the usefulness of the Carnot engine efficiency is the efficiency of a steam power plant. For a steam power plant, the high temperature source temperature is the maximum temperature of the steam in the boiler. The low temperature source temperature is the temperature of the outlet water in the condenser. If $T_{low} = 50°F$ (509.69°R) and $T_{high} = 1000°F$ (1459.69°R), the maximum efficiency is equal to $(1-509.69/1459.69)$ = 65%. If it was not for this maximum efficiency, one would think that the maximum efficiency could be 100%, and the actual efficiency of a (quite efficient) steam power plant, say 40%, would seem extremely low.

   A Stirling engine is similar to a Carnot. It is an external combustion engine but unlike the Carnot engine the Stirling engine employs heat regeneration within the engine's cylinder. The regeneration

---

*See Appendix for derivation
†°R = °F + 459.69

device absorbs heat from the gas that is within the engine cylinder during part of the cycle, and rejects heat to the gas at another part of the cycle. The Stirling engine has the same efficiency as a Carnot engine. (See the appendix for the derivation of the Stirling engine efficiency). The Stirling engine was first suggested by Reverend Robert Stirling (1797-1878).

A diagram of a Stirling engine is shown in Figure 11-2. Note that there are several different arrangements of Stirling engines and the engine shown in Figure 11-2 is intended to make it easier to understand the theory of the Stirling engine. The diagram in Figure 11-2 is from Reference 22. Note that both the Carnot and the Stirling engines transfer heat at constant temperature only with regard to the high and low temperature sources.

The efficiency of a Stirling solar system is about 30%. This efficiency is defined as the electrical power to the grid divided by the incoming solar energy at the mirrors. Using this efficiency value and the file *Two Axis Tracking* from the folder *Incident — Absorbed Solar Radiation*, one can estimate the electrical output of a Stirling engine system for locations in the USA. As an example, consider a Stirling engine system in the Tucson, Arizona area which has ten dishes with an area of 908 ft^2 per mirror. What is the yearly average output of the system?

The output is 753,000 kilowatt-hours; this is enough energy to supply 67 homes.

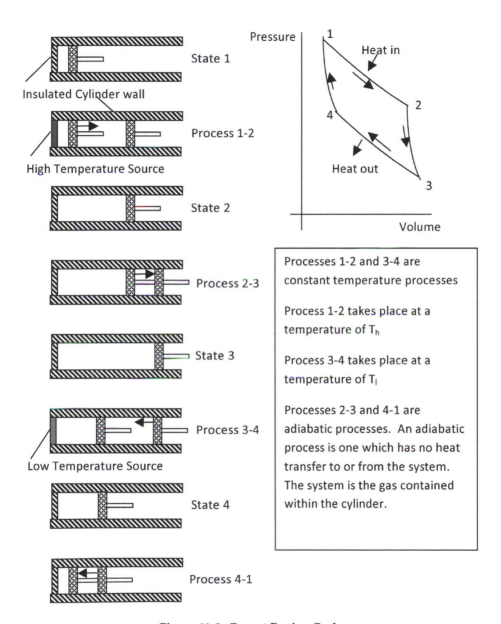

Figure 11-2. Carnot Engine Cycle

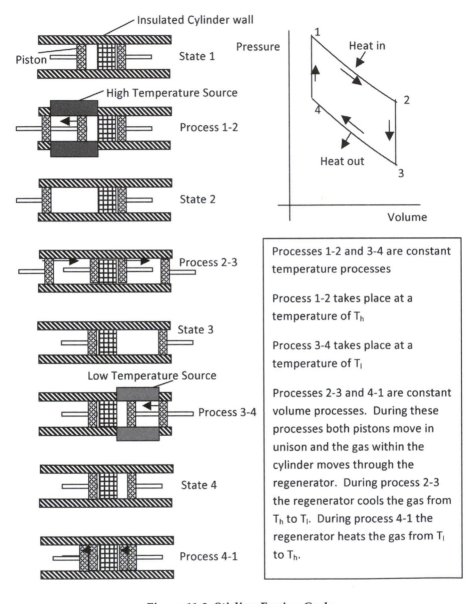

Figure 11-3. Stirling Engine Cycle

Chapter 12

Passive Solar Energy

Up to this point, all the solar energy systems we have discussed have been active systems. Active systems utilize machinery or photovoltaic elements. Passive solar energy systems generally have no moving parts and typically are made up of windows and sometimes walls.

Windows placed on the south side of a home are passive solar elements. In the winter months, sunshine enters the windows on a sunny day and heats the room. On cloudy days and at night, heat is lost out of the window. In this chapter, a program will be developed which estimates the net heat transfer from windows. This program accounts for closing window shades on cloudy days and at night. Before describing the passive window program, consider the shading of windows to keep out some of the summer sun. Window shading is shown in Figure 12-1.

The angles θ_S and θ_W can be determined using the program on the file *Sun Location*. In using the sun location program, the day of the year is needed. Use Table 4-1 in Chapter 4 to determine the day of the year. Referring to Figure 12-1 we have:

$$\tan \theta_S = (W + H)/L$$

$$\tan \theta_W = H/L$$

Solving these equations for L & H:

$$L = W/(\tan\theta_S - \tan\theta_W) \tag{12-1}$$

$$H = L \tan\theta_W \tag{12-2}$$

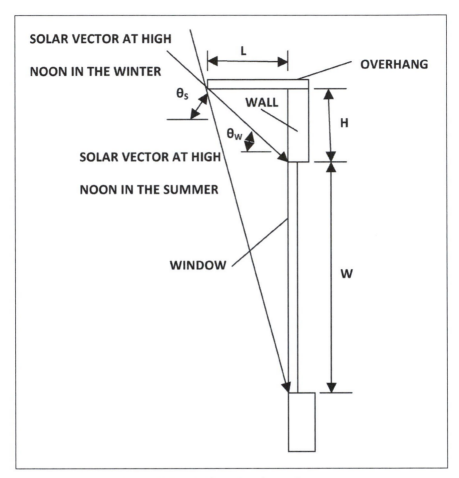

Figure 12-1. Window Overhang Geometry

To illustrate the use of the equations given above, assume a location with a latitude of 40 degrees and a window height of 5 ft. Also, assume the window will be totally shaded at noon on the 15th of August and the sun's shadow will be at the top of the window on the 15th of February. The day of the year for February 15th is 46 and for August 15th is 227. Using the file, *Sun Location*, set the tilt angle to 90° and the azimuth angle to 0° (vertical window facing due south). For high noon on the days given above, $\theta_S = 64.3°$ and $\theta_W = 37.05°$. From equation (62) L (the overhang) = 3.78 ft and from equation (63) H (distance from top of window to the overhang) =

2.85 ft. Note that on August 15th the window will be totally shaded at noon but the window will be only partially shaded before and after noon.

Earlier in this chapter, a wall was mentioned as a passive solar energy device. In Figure 12-2, a Trombe wall is shown. This device is credited to the French engineer, Felix Trombe, who was, in the 1980's, the director of the French solar facility in the Pyrenees. Incidentally, the author met Trombe at the University of Miami in the 1980's. He was a very distinguished gentleman resembling President de Gaulle.

The wall serves as a heat storage device. When the sun shines, the wall is heated, and at night or during cloudy periods heat that was stored in the wall is transferred to the room. The material used to construct the wall needs to have a relatively high heat capacity given by the product of its density times its specific heat. This quantity is shown in Table 12-1 for various materials per unit volume. Along with this quantity, the radiation absorptivity of the wall surface receiving the solar radiation should be high. A dark color along with a fairly rough surface will generally provide a high absorptivity.

A thermal model of a Trombe wall is shown in Figure 12-3. The heat input, Q, is consistent with the average incident radiation for Denver, Colorado times 0.72 to account for losses for transmission through two pains of glass and the absorbtivity on the wall surface. The incident radiation versus time is shown in Figure 12-4. Temperature gradients within the wall are shown in Figure 12-5.

W = wall thickness, ft
A = area of wall perpendicular to the heat flow direction, ft^2
k = thermal conductivity of wall material, Btu/(hr-ft-°F)
Cp = specific heat of wall material, Btu/(lb-°F)
ΔX = W/5, ft
M = $A^*W/5^*\rho^*Cp$, Btu/°F
Q = Incident solar radiation absorbed by wall, Btu/hr
h_1 = free convection film coefficient at left face of wall, Btu/(hr-ft^2-°F)

h_2 = free convection film coefficient at right face of wall, Btu / (hr-ft^2-°F)

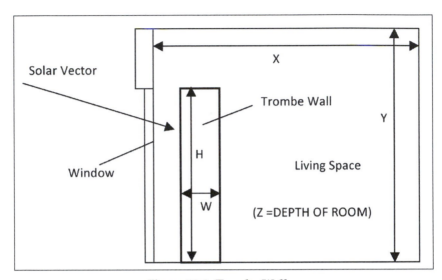

Figure 12-2. Trombe Wall

Table 12-1. Heat Capacity per Unit Volume

MATL	DENSITY, ρ LB/FT^3	SPECIFIC HEAT, BTU/LB-°F	HEAT CAPACITY, BTU/FT^3-°F
BRICKS	165	0.23	38.0
CONCRETE	130	0.21	27.3
GRANITE	165	0.195	32.2
FIR	26	0.65	16.9
PINE	40	0.67	26.8
SAND	94.6	0.191	18.1
ALUMINUM	174	0.211	36.7
IRON	493	0.1737	85.6
WATER	62.4	1	62.4

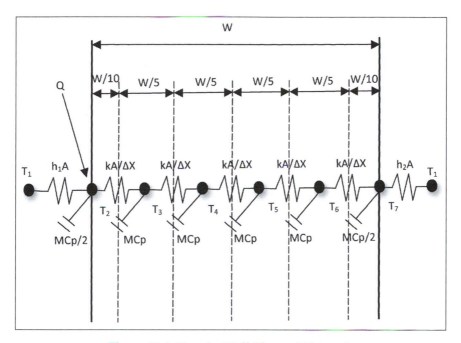

Figure 12-3. Trombe Wall Thermal Network

The curve labeled Maximum Temperatures in Figure 12-5 occurred at hour 38.33 and the curve labeled Temperature at Sunup occurred at time 30.78. These times are consistent with those shown in Figure 12-4. The wall material used for the analysis shown above was bricks. It was assumed that the room air on both sides of the wall were constant at 70°F.

When passive solar windows are used, there is a heat gain during periods of sunlight. On the other hand, heat is lost at night and during cloudy periods during the day. A program has been devised which calculates the net gain or loss for windows with double-glazing. This program is on the file 'Passive-Windows'. When using this program, the tilt and azimuth angles of the window can be defined. The analysis can be performed for any of the more than 150 US cities listed on the files *Weather Data-F* and *Weather Data-TT*. Note that data from both of these files are needed to perform the Passive-Window analyses. Other required input data to the Passive-Window program are the overall heat transfer coefficients (U) for the window and for the window with drawn shades. For an example, in Denver, Colorado, without using shades the gain is 1.35E5 Btu / (sq ft-year) and with shades used when sun is not shining the gain is 1.903E5 Btu / (sq ft-year). The equivalent values for Seattle, Washington, are 6.67E4 and 1.18E5.

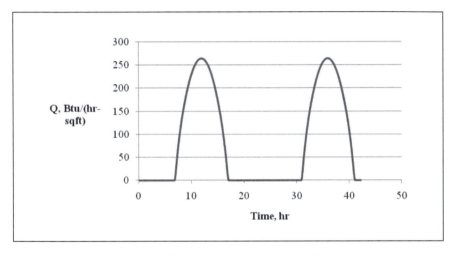

Figure 12-4. Incident Solar Radiation

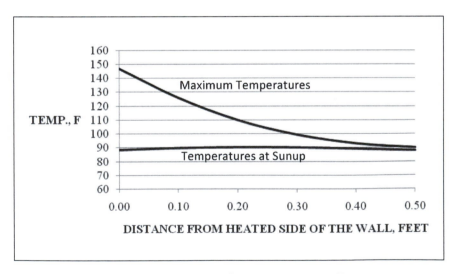

Figure 12-5. Trombe Wall Temperature Gradients

Chapter 13

Greenhouse Solar Collector

A plan view of the author's home is shown in Figure 13-1. The greenhouse is used for growing vegetables and for providing solar heat to the house. The greenhouse is 8 x 16 ft with a double pane transparent roof and two double pane windows on the south wall. The window dimensions are 6'8" x 2'. The construction is a rock wall two ft above ground level with 2 x 6" framing for the rest of the structure. Fiberglass insulation is used in the walls. The exterior of the wall is ¾" cedar and the interior is ¾" pine. Approximately 50 one-gallon plastic containers of water are stored in the greenhouse to provide thermal storage.

A diagram of the active heating system is shown in Figure 13-2. The two thermostats are set at 70°F. If the temperature at thermostat H is less than 70°F and the temperature at thermostat

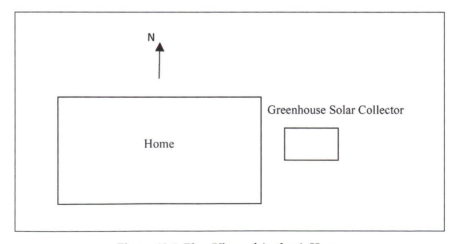

Figure 13-1. Plan View of Author's Home

G is greater than 70°F, the circuit is closed and the fans come on. If the temperature of the greenhouse is greater than 120°F, the fan in the greenhouse comes on and ventilates the greenhouse. There are two louvered windows on the greenhouse wall opposite the fan and a single louvered window at the fan. These windows automatically open when the greenhouse fan comes on. This is common practice for greenhouse ventilation.

Along with the greenhouse solar collector system, the home is outfitted with passive solar windows. There is a total of 208 square ft of double pane windows on the south side of the home.

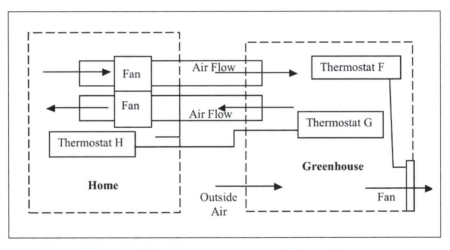

Figure 13-2. Diagram of Greenhouse Solar Collector

Problems

Problems are provided to help the reader apply knowledge gained from the book to real-world examples. Solutions to these problems are provided in the companion CD.

1. Convert 40°C to (a) °F and (b) °R.

2. Determine the declination for (a) September 12th and (b) December 29th.

3. What are sunup and sundown times for (a) June 21st and (b) October 5th in Seattle, Washington?

4. What is the average monthly incident radiation in Btu/ft^2 on a horizontal surface for the month of July in Las Vegas, Nevada?

5. A spherical tank (4 foot outside diameter) is insulated with a layer of glass wool (4 lb/ft^3) four inches thick. The tank wall is at a temperature of -120°F and the outer surface of the insulation is at 80°F. What is the heat loss from the tank in Btu/hr?

6. A vertical surface four ft high is maintained at a temperature of 100°F. Room air at 70°F is adjacent to the wall. What is the Grashof number for this situation?

7. Water flows in a tube with a 1.5 inch inside diameter. The tube wall is maintained at a constant temperature of 200°F and the inlet water temperature is 100°F. The water flow rate is 10 gal/min. Note that 7.48 gal = 1 ft^3. If the tube is 20 ft long what is the outlet temperature of the fluid?

8. Two surfaces are positioned in space as shown below. The dimensions are in ft.

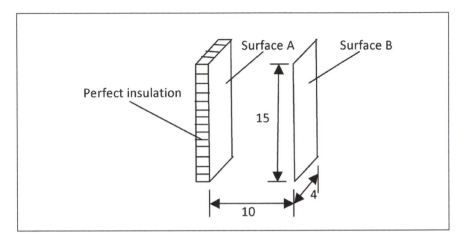

Surface B is maintained at a constant temperature of 500°F and surface A is perfectly insulated on the back side. Both surfaces are 4 x 15 ft and directly opposed to each other. The IR emissivities of both surfaces are 0.8. What is the temperature, in °F, of surface A?

9. The properties of a flat plate solar collector with two cover plates are shown below.

ABSORBER AREA, FT^2	24
GAP-1-ABSORBER T0 GLASS-2, IN	1.20
GAP- 2- GLASS-2 TO GLASS-1, IN	1.0
AREA OF BACK INSULATION, FT^2,	24
BACK INSULATION THICKNESS, IN	1
INSULATION CONDUCTIVITY, Btu / ((HR-FT-F)	0.02
EDGE INSULATION HEIGHT, IN	3.2
EDGE INSULATION THICKNESS, IN	1
EDGE INSULATION TOTAL LENGTH, FT	22
INLET FLUID TEMPERATURE, °F	50
PERIMETER OF ABSORBER FLUID CHANNEL, IN	1.5
AREA OF ABSORBER FLUID CHANNEL, IN^2	0.099
HYDRAULIC DIA. OF ABSORBER FLUID CHANNEL, IN	0.264
# OF PARALLEL ABSORBER FLUID CHANNELS	8
LENGTH OF ABSORBER FLUID CHANNELS, FT	64
AVE ABSORBER FLUID CHANNEL SPACING, IN	4.5
THICKNESS OF ABSORBER PLATE, IN	0.09375

K OF ABSORBER MATERIAL, Btu/(HR-FT-°F)	128
ABSORBER IR EMISSIVITY	0.85
ABSORBER REFLECTIVITY	0.15
ABSORBER SOLAR ABSORBTIVITY	0.87
ABSORBER SOLAR REFLECTIVITY	0.13
GLASS-1 IR EMISSIVITY	0.94
GLASS-1 IR REFLECTIVITY	0.03
GLASS -1 IR TRANSMISSIVITY	0.03
GLASS -1 SOLAR ABSORBTIVITY	0.01
GLASS -1 SOLAR REFLECTIVITY	0.095
GLASS -1 SOLAR TRANSMISSIVITY	0.895
GLASS-2 IR EMISSIVITY	0.94
GLASS-2 IR REFLECTIVITY	0.03
GLASS -2 IR TRANSMISSIVITY	0.03
GLASS -2 SOLAR ABSORBTIVITY	0.01
GLASS -2 SOLAR REFLECTIVITY	0.095
GLASS -2 SOLAR TRANSMISSIVITY	0.895
AMBIENT TEMPERATURE, °F	50
ELEVATION, FT	0
WIND VELOCITY, MPH	10
INCOMING SOLAR RADIATION, Btu/(HR-FT^2)	300
ANGLE OF INCIDENCE, DEGREES	0
FLUID VELOCITY, FT/SEC	3
COLLECTOR TILT ANGLE, DEGREES	42

Vary the fluid inlet temperature and the incoming solar radiation to calculate collector efficiencies for different values of $(T_i - T_a)/I$. T_i is the inlet temperature, T_a is the ambient temperature, and I is the incident solar energy on the collector surface. Use these values and the Trendline feature of Microsoft Excel to determine the efficiency equation for the collector defined by the data given above.

10. Find the yearly absorbed solar radiation for a 30 ft^2 flat plate collector located in Duluth, Minnesota. The collector's $F'\tau\alpha = 0.7$ and $F'UL = 0.8$ Btu/(hr-ft^2-°F). The azimuth angle of the collector $= 20°$ and the inlet fluid temperature $= 80°$F. The wind speed is 10 mph.

11. Find the payback period for a solar domestic hot water heating system in Austin, Texas. The number of people in the household is five and the fuel used for water heating is natural gas costing $14.5/1000 ft³. The basic system cost is $3000 with one 24 ft² collector. Each additional collector costs $950. The collector's F'τα = 0.75 and the F'UL = .85. The azimuth angle is 0 and the inlet water temperature is 50°F.

12. A photovoltaic system's overall efficiency (incoming solar radiation to output electricity) is 11.5 %. The system is to have an output of 1000 MW-hr/yr (1 Btu = 3.412 watt-hr). Assume that the system will be a one axis azimuth tracking system. Find the optimum tilt angle and the total array area to yield the 1000 MW-hr/yr required system output. Assume the system is located in Yuma, Arizona.

13. The wall of a frame building is made of pine 2 x 4's on 16 inch centers. The outside of the wall is made up of ¾ inch plywood under ¾ inch cypress siding. The inside is ½ inch drywall. Assume the outside film coefficient is 10 Btu/(hr- ft²-°F) and the inside film coefficient is 0.5 Btu/(hr- ft²-°F). The 5 ½ inch space between the inside and outside wall panels is filled with 4 lb/cu ft glass wool. What is the overall heat transfer coefficient (U) for the wall?

14. A home in Helena, Montana, has a total overall heat transfer coefficient times effective area (UA) of 423 Btu/(hr-°F). The basic solar heating system cost with one collector is $3550. Each additional 30 ft² collector costs $1050. The collector's F'τα = 68% and the F'UL = 0.8. Assume the home is heated by electricity at a cost of 0.18 $/(kW-hr). Find the payback period for the system if the total incentive for installing the system is $1200. Assume the inside temperature is 70°F.

15. Consider the two heliostats shown in the figure below. Find the incident and reflected angles for heliostats A and B.

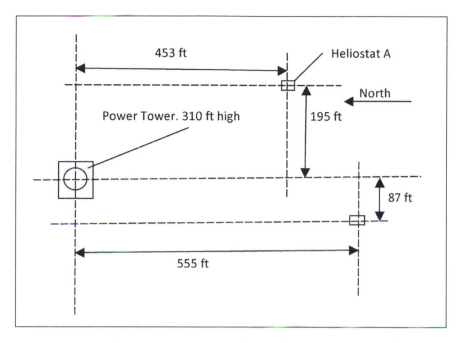

Assume the system is located in Albuquerque, New Mexico, the date is December, 30th, and the time of day is 11:45 am.

16. A Stirling engine is powered by a heat source at a temperature 1250°F. The cooling provided for the engine is maintained at a temperature of 190°F. What is the maximum possible efficiency for this engine?

17. A passive solar window is shown below. The window is 4½ ft high (W).

   The location of the window is Miami, Florida. Find the length of the overhang (L) and the height of the overhang above the window (H) to satisfy the following conditions. At noon on June 21st, the overhang just shadows the bottom of the window, and noon on December 21st, the shadow line is at the top of the window.

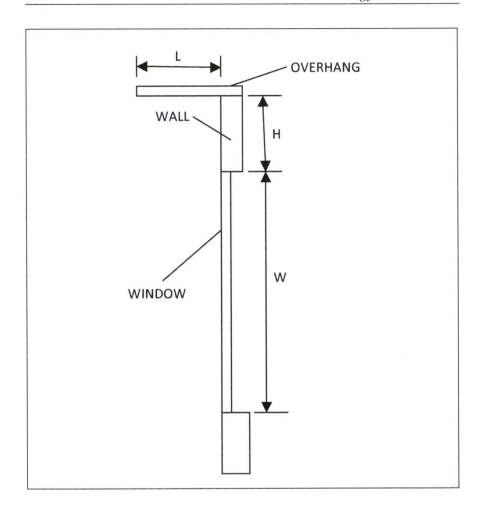

References

Duffie, John A. and Beckman, William A., *Solar Energy Thermal Processes*, John Wiley & Sons,1974.

1999 Applications Handbook of the American Society of Heating, Refrigeration, and Air-Conditioning Engineers

Spencer, J.W., Fourier series representation of the position of the Sun, 2(5), 172.

Hansen, H., "Darstellung des Warmeuberganges in Rohren durch verallgemeinerte Potenzbeziehungen," *Zeitschr. V.D.I. Beihefte Verfahrenstechnik*, No. 4, 1943, p 91.

McAdams, *Heat Transmission*, McGraw-Hill, 3rd Ed., 1954.

Siegel, R., Howell, J.R., *Thermal Radiation Heat Transfer*, Vol 1, NASA Lewis Research Center, Cleveland, OH, 1968.

Oppenheim, A.K., "Radiation Analysis by the Network Method," Transactions of the ASME, Vol. 78, 1956, p. 725.

Chapman, A.J., *Heat Transfer*, 2nd Ed., MacMillian, New York, 1960.

Kreith, *Principles of Heat Transfer*, International Textbook Company, 2nd Ed., 1965.

Holman, *Heat Transfer*, McGraw-Hill, 2nd Ed., 1968.

Schneider, *Temperature Response Charts*, John Wiley and Sons, Inc., 1963.

Arpaci, *Conduction Heat Transfer*, Addison-Wesley, 1966.

Patankar, *Numerical Heat Transfer and Fluid Flow*, McGraw-Hill, 1980.

Weibelt, *Engineering Radiation Hear Transfer*, Holt, Rinehart, & Winston, 1965.

Kreith, *Radiation Heat Transfer for Spacecraft and Solar Power Plant Design*, International Textbook Co., 1962

Sparrow and Cess, *Radiation Heat Transfer*, Brooks/Cole Publishing Company, 1966.

McMordie, R.K., *Analytical Model of Flat Plate Solar Collectors*, Proceedings of the Solar Cooling and Heating Forum, December 13-15, 1976, Miami Beach. Florida, p 195.

Buchberg, H, Catton, I and Edwards, D.K., *Natural Convection in*

Enclosed Spaces — A Review of Application to Solar Energy Collection, Transactions of the ASME, May 1976.

Aldrich, Robb & Vijayakumar, Gayathri, *Cost, Design, and Performance of Solar Hot Water in Cold-Climate Homes, Steven Winter Associates, Inc.*

1985 Fundamentals Handbook of the American Society of Heating, Refrigeration, and Air-Conditioning Engineers

McMordie, R. K., *Journal of Solar Energy Engineering,* Transactions of the ASME, February 1984, p 98.

Jones, J.B. and Hawkins, G.A. *Engineering Thermodynamics*, John Wiley & Sons, Inc, 1960.

Appendices

Appendix A

Carnot and Stirling Efficiencies

CARNOT EFFICIENCY

The Carnot engine cycle is made up of two constant temperature and two adiabatic processes. It assumed that the gas in the engine cylinder is an ideal gas. For an ideal gas we can write:

$$Pv = RT \tag{1}$$

Where P is pressure, v is specific volume (volume/mass), R is the ideal gas constant, and T is absolute temperature (°R or °K)

For an ideal gas and an adiabatic process:

$$Pv^k = c = \text{constant} \tag{2}$$

Where k is the ratio of specific heat at constant pressure (C_p) to specific heat at constant volume (C_v).

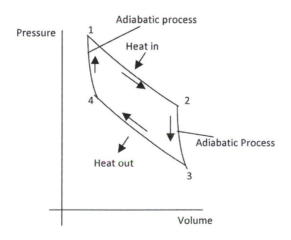

111

The pressure/volume diagram given below shows the Carnot cycle.

Referring to the diagram the net work output of the cycle is given by:

$$W_{cycle} = {}_1W_2 + {}_2W_3 + {}_3W_4 + {}_4W_1$$

$${}_1W_2 = \int_1^2 Pdv \tag{3}$$

From equation (1):

$$P = RT/v \tag{4}$$

Substituting equation (4) into equation (3):

$${}_1W_2 = RT\int_1^2 \frac{dv}{v} = RT_1 \ln(v_2/v_1) \tag{5}$$

$${}_2W_3 = \int_2^3 Pdv \tag{6}$$

Substituting equation (2) into equation (6):

$${}_2W_3 = c\int_2^3 dv/v^k = cv_3^{(1-k)}/(1\!-\!k)\!-\!cv_2^{(1-k)}/(1\text{-}k)$$

$$=[c/(1\text{-}k)](v_3^{(1\text{-}k)}\!-\!v_2^{(1\text{-}k)}) \tag{7}$$

Working with equation (2):

$$c = P_2v_2^k = P_3v_3^k$$

Substituting these results into equation (7):

$${}_2W_3 = [1/(1\text{-}k)]\,(P_3v_3^k\,v_3^{(1\text{-}k)}\!-\!P_2v_2^k\,v_2^{(1\text{-}k)})$$
$$= (P_3v_3\!-\!P_2v_2)/(1\text{-}k) \tag{8}$$

Substituting equation (1) into equation (8):

$$_2W_3 = R(T_3-T_2)/(1-k)$$

Using the same approach as above:

$$_3W_4 = RT_3 \ln(v_4/v_3) \tag{9}$$

$$_1W_4 = R (T_1—T_4)/(1-k) \tag{10}$$

We can now write:

$$W_{cycle} = RT_1 \ln(v_2/v_1)+R(T_3-T_2)/(1-k)$$
$$+ RT_3 \ln(v_4/v_3)+ R (T_1—T_4)/(1-k) \tag{11}$$

Recognizing that $T_1 = T_2 = T_{high}$ and $T_3 = T_4 = T_{low}$ equation (11) can be rewritten:

$$W_{cycle} = RT_{high} \ln(v_2/v_1)+R(T_{low}-T_{high})/(1-k)$$
$$+ RT_{low} \ln(v_4/v_3)+ R (T_{high}—T_{low})/(1-k)$$

$$W_{cycle} = RT_{high} \ln(v_2/v_1)+ RT_{low} \ln(v_4/v_3)$$

The heat input to the cycle is:

$$_1Q_2 = {}_1W_2 = = RT_{high} \ln(v_2/v_1)$$

The cycle efficiency is:

$$\eta = (\text{work out})/(\text{heat in}) = W_{cycle}/{}_1Q_2$$
$$= [RT_{high} \ln(v_2/v_1)+ RT_{low} \ln(v_4/v_3)]/RT_{high} \ln(v_2/v_1)$$
$$= [1 + T_{low} \ln(v_4/v_3)/T_{high} \ln(v_2/v_1)]$$
$$= \{1 +(T_{low}/T_{high}) [\ln(v_4/v_3)/\ln(v_2/v_1)]\} \tag{12}$$

Noting that:

$$\ln(v_4/v_3 = - \ln(v_3/v_4)$$

Rewrite equation (22):

$$\eta = \{1 - (T_{low}/T_{high}) [\ln(v_3/v_4)/\ln(v_2/v_1)]\} \tag{13}$$

Working with the specific heats:

$$P_2 v_2 = RT_2 \text{ Ideal gas equation of state} \tag{14}$$

$$P_2 v^k = c \text{ (constant) defines reversible adiabatic process} \tag{15}$$

Solve for P in equation (15) and substitute into equation (14):

$$c\, v_2/v_2^k = RT_2$$

Rearranging:

$$v_2 = \{(R/c)\, T_2\}^{\wedge}[1/(1-k)] \tag{16}$$

The equivalent equation for v_3 is:

$$v_3 = \{(R/c)\, T_3\}^{\wedge}[1/(1-k)] \tag{17}$$

Dividing equation (16) by equation (17):

$$v_2/v_3 = (T_2/T_3)^{\wedge}[1/(1-k)] = (T_{high}/T_{low})^{\wedge}[1/(1-k)] \tag{18}$$

Repeating the process, starting with equation (14), but using v_1 and v_4, we have:

$$v_1/v_4 = (T_1/T_4)^{\wedge}[1/(1-k)] = (T_{high}/T_{low})^{\wedge}[1/(1-k)] \tag{19}$$

Equating equations (18) and (19):

$$v_2/v_3 = v_1/v_4$$

$$v_3/v_4 = v_2/v_1 \tag{20}$$

Substituting equation (20) into equation (13), we have:

$$\eta = 1 - T_{low} / T_{high}$$

STIRLING EFFICIENCY

The Stirling engine cycle is made up of two constant temperature and two constant volume processes. It assumed that the gas in the engine cylinder is an ideal gas. As for the Carnot efficiency, we start with the ideal gas equation:

$$Pv = RT \qquad (1)$$

With P, v, R & T as previously defined.

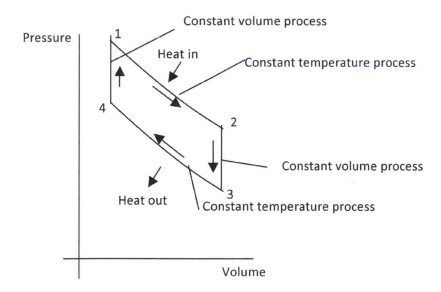

Referring to the diagram, the net work output of the cycle is given by:

$$W_{cycle} = {}_1W_2 + {}_2W_3 + {}_3W_4 + {}_4W_1$$

$$_1W_2 = \int_1^2 Pdv \tag{2}$$

From equation (1):

$$P = RT/v \tag{3}$$

Substituting equation (3) into equation (2):

$$_1W_2 = RT\int_1^2 \frac{dv}{v} = RT_1 \ln(v_2/v_1) \tag{4}$$

$$_2W_3 = \int_2^3 Pdv \tag{5}$$

Process 2 to 3 is a constant volume process, therefore:

$$_2W_3 = 0 \tag{6}$$

Using the same approach as above:

$$_3W_4 = RT_3 \ln(v_4/v_3) \tag{7}$$

$$_1W_4 = 0 \tag{8}$$

We can now write:

$$W_{cycle} = RT_1 \ln(v_2/v_1) + RT_3 \ln(v_4/v_3) \tag{9}$$

Recognizing that $T_1 = T_2 = T_{high}$ and $T_3 = T_4 = T_{low}$ equation (9) can be rewritten:

$$W_{cycle} = RT_{high} \ln(v_2/v_1) + RT_{low} \ln(v_4/v_3)$$

The heat input to the cycle is:

$$_1Q_2 = {}_1W_2 = = RT_{high} \ln(v_2/v_1)$$

Note that there is no heat input to the engine during processes 2 to 3 or 4 to 1. Heat transfer during these processes is internal to the engine using the regenerator.

The cycle efficiency is:

$$
\begin{aligned}
\eta \quad &= (\text{work out})/(\text{heat in}) = W_{cycle}/{}_1Q_2 \\
&= [\, RT_{high} \ln(v_2/v_1) + RT_{low} \ln(v_4/v_3)\,]/RT_{high} \ln(v_2/v_1) \\
&= [1 + T_{low} \ln(v_4/v_3)/T_{high} \ln(v_2/v_1)] \\
&= \{1 + (T_{low}/T_{high}) \, [\ln(v_4/v_3)/\ln(v_2/v_1)]\}
\end{aligned}
\tag{10}
$$

Noting that:

$$\ln(v_4/v_3) = -\ln(v_3/v_4)$$

Rewrite equation (22):

$$\eta = \{1 - (T_{low}/T_{high}) \, [\ln(v_3/v_4)/\ln(v_2/v_1)]\} \tag{11}$$

Noting that $v_2 = v_3$ and $v_1 = v_4$, we have:

$$\eta = 1 - T_{low}/T_{high} \tag{12}$$

This efficiency equation is identical to the efficiency equation for the Carnot engine.

Appendix B

Mathematical Techniques for Solving Heat Transfer Problems

INTRODUCTION

There are two basic mathematical approaches for solving heat transfer problems. These are the analytical approach and the numerical approach. With the analytical approach the mathematical expression describing the heat transfer problem is solved resulting in an exact equation. Using the numerical approach the physical characteristics of the heat transfer problem are broken down into small, but finite, pieces. These finite pieces are called nodes and are solved individually using relatively simple equations.

It is my judgment that most heat transfer problems encountered in industry are best solved using numerical approaches. Many problems cannot be solved analytically because of the complex geometry involved in the problem or due to thermal properties being dependent on temperature. These characteristics cause analytical solutions to be very difficult or, indeed, often impossible to derive.

There are occasions when analytical solutions can be determined, which approximate the actual problem. For example, an analytical solution may be found which fits the actual problem except the thermal properties cannot be treated as functions of temperature. For this situation it is advantageous to use the analytical solution with constant properties. Develop a numerical solution with constant properties to match the analytical solution. Compare these solutions and adjust the numerical solution until there is a close correspondence between the approaches. For example, the nodal breakdown of the numerical model may have to be adjusted to allow a close match between the analytical and numerical results. When the solution of the numerical model closely matches the analytical model, then the numerical model can be modified to allow for properties to be functions of temperature or other adjustments so that the numerical model simulates the actual heat transfer hardware as closely as possible.

ANALYTICAL SOLUTION— STEADY-STATE WITH CONDUCTION AND CONVECTION

Consider the heat transfer fin, shown in Figure B-1 and assume steady-state conditions.

In Figure B-1, heat in Btu/hr is being transferred along the fin from left to right. The left base of the fin is maintained at a temperature of T_b in °F. The temperature of the fluid surrounding the fin is T_a in °F. Heat is lost to the surrounding fluid along the fin and at the end of the fin. The film coefficient along the fin is h and the film coefficient at the right end of the fin is h_{end}. The units of film coefficient are Btu/(hr-ft2-°F). The thermal con-

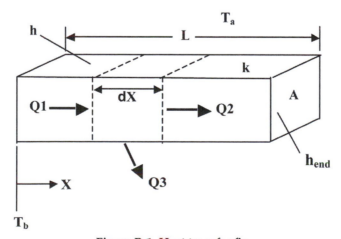

Figure B-1. Heat transfer fin

ductivity of the fin material is k in Btu/(hr-ft-°F). A is the cross sectional area of the fin in ft2 and C (not shown in the figure) is the circumference of the fin's cross section in ft. The analytical solution for the temperature along the fin is given by equation (B-1) while the heat loss by the fin is given by equation (B-2).

$$T = T_a + (T_b - T_a)\cosh(\beta X) - \frac{(T_b - T_a)\left(\frac{h_{end}}{\beta}\cosh(\beta L) + k\sinh(\beta L)\right)\sinh(\beta X)}{\frac{h_{end}}{\beta}\sinh(\beta L) + k\cosh(\beta L)}$$

(B-1)

$$Q = kA\frac{(T_b - T_a)\beta\left(\frac{h_{end}}{\beta}\cosh(\beta L) + k\sinh(\beta L)\right)}{\frac{h_{end}}{\beta}\sinh(\beta L) + k\cosh(\beta L)}$$

(B-2)

where,

$$\beta = \sqrt{\frac{hC}{kA}}$$

h = the film coefficient along the fin in Btu/(hr-ft2-°F)

h_{end} = the film coefficient at the end of the fin in Btu/(hr-ft2-°F)

C = the circumference of the cross section of the fin in ft

k = the thermal conductivity of the fin material in Btu/(hr-ft-°F)

A = the area of the fin's cross sectional in ft2

NUMERICAL SOLUTION—
STEADY-STATE WITH CONDUCTION AND CONVECTION

We will again consider a heat transfer fin, but will use a numerical approach to solve the steady-state problem.

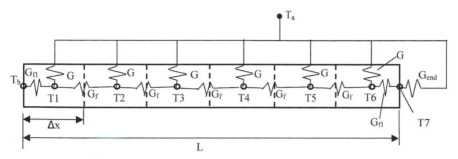

Figure B-2. Heat transfer fin, numerical approach

In Figure B-2 the fin has been divided into six pieces or nodes. Additionally, there are node points at the base of the fin, T_b, at the end of the fin, T7, and representing the ambient air, T_a. Thermal conductors are connected between the nodes, as shown in Figure B-2. The conductor values are as follows:

$$
\begin{aligned}
G &= h\ As \\
G_f &= k\ A/\Delta x \\
G_{f1} &= k\ A/(\Delta x/2 \\
G_{end} &= h_{end}\ A
\end{aligned}
$$

where:

h = the film coefficient along the fin in Btu/(hr-ft2-°F)

h_{end} = the film coefficient at the end of the fin in Btu/(hr-ft2-°F)

k = the thermal conductivity of the fin material in Btu/(hr-ft-°F)

A = the area of the fin's cross sectional in ft2

A_s = surface area of the nodes in ft2

Δx = length of the node in ft

We will compare the analytical solution of the fin problem to a numerical solution. Assume the following characteristics of the problem:

h = 1.7611 Btu/(hr-ft2-°F)

$$
\begin{aligned}
h_{end} &= 2.6417 \text{ Btu}/(\text{hr-ft}^2\text{-}^\circ\text{F}) \\
k &= 115.56 \text{ Btu}/(\text{hr-ft-}^\circ\text{F}) \\
A &= 4.306\text{E-}03 \text{ ft}^2 \\
A_s &= 4.306\text{E-}02 \text{ ft}^2 \\
\Delta x &= 1.640\text{E-}01 \text{ ft} \\
L &= 9.843\text{E-}01 \text{ ft} \\
T_b &= 440.31 \ ^\circ\text{F} \\
T_a &= 80.31 \ ^\circ\text{F}
\end{aligned}
$$

Using equations (B-1) and (B-2) we obtain the analytical, or exact solution:

$$
\begin{aligned}
T7 &= 318.59 \ ^\circ\text{F} \\
Q &= 129.457 \text{ Btu/hr}
\end{aligned}
$$

The numerical solution is derived by performing energy balances at each of the node points, except for the boundary nodes, Tb and Ta. An energy balance at node T1 yields:

$$G_{fi}(Tb - T1) + G(Ta - T1) + G_f(T2 - T1) = 0$$

Solving for T1:

$$T1 = (G_{f1} \ Tb + G \ T_a + G_f \ T2)/(G_{f1} + G + G_f) \tag{B-3}$$

An energy balance at node T2 yields:

$$G_f(T1 - T2) + G(T_a - T2) + G_f(T3 - T2) = 0$$

Solving for T2:

$$T2 = (G_f \ T1 + G \ T_a + G_f \ T3)/(2G_f + G) \tag{B-4}$$

This process is repeated to give the following equations:

$$T3 = (G_f \ T2 + G \ T_a + G_f \ T4)/(2G_f + G) \tag{B-5}$$

$$T4 = (G_f \, T3 + G \, T_a + G_f \, T5)/(2G_f + G) \qquad (B\text{-}6)$$

$$T5 = (G_f \, T4 + G \, T_a + G_f \, T6)/(2G_f + G) \qquad (B\text{-}7)$$

$$T6 = (G_f \, T5 + G \, T_a + G_{f1} \, T7)/(G_f + G_{f1} + G) \qquad (B\text{-}8)$$

$$T7 = (G_{f1}T6 + G_{end}T_a)/(G_{f1} + G_{end}) \qquad (B\text{-}9)$$

Equations (B-3) through (B-9) are a set of seven linear equations with seven unknowns. Solving these equations simultaneously gives a value for T7 of 318.76°F. This is the value of the numerical solution for node T7 and is within 0.134% of the analytical solution from equation (B-1).

Another numerical technique for solving the seven linear equations is relaxation. The first step in solving a set of equations by relaxation is to guess values for the unknown temperatures. Then apply the equations (B-3) through (B-9) to calculate new values for the guessed temperatures. This step represents the first iteration of the problem. Repeat the application of the linear equations to calculate a new set of temperatures for a second iteration of the problem. This process is repeated over and over until the temperatures converge to constant values. The relaxation process for the fin problem is illustrated in Table B-1.

The heat transfer from the fin is calculated using the numerical approach by the following equation:

$$Q = k \, A(T_b - T1)/(\Delta x/2) = 115.56 * 4.306E\text{-}03*(440.31 -$$
$$419.05)/(1.6404E\text{-}01/2) = 129.08 \text{ Btu/hr}$$

This numerical value is within 0.291% of the exact value given by equation (B-2). The accuracy of the numerical solution can be improved by dividing the fin into smaller nodes, that is, to increase the number of nodes. If the number of nodes is doubled to 12, the percent difference in the fin end temperature between the analytical solution and numerical solution is 0.0046%. The difference for the heat transfer from the fin is 0.105% between the analytical and numerical solutions.

Table B-1. Relaxation solution for the steady-state fin problem

ITERATION	T1	T2	T3	T4	T5	T6	T7
1	260.31	260.31	260.31	260.31	260.31	260.31	260.31
2	377.830	258.087	258.087	258.087	258.087	258.822	259.973
3	377.095	315.025	255.892	255.892	256.255	257.865	258.488
4	395.918	313.578	282.926	253.904	254.699	256.277	257.532
5	395.44	336.223	281.229	266.485	252.933	255.131	255.948
..							
..							
300	419.047	384.991	358.553	339.071	326.0599	319.1917	318.7449
..							
..							
452	419.0487	384.9948	358.5582	339.078	326.0672	319.2006	318.7535

ANALYTICAL SOLUTION—
TRANSIENT WITH CONDUCTION AND CONVECTION

The expression for transient heat transfer in one dimensional, rectangular coordinates is:

$$\frac{\partial T}{\partial t} = \frac{k}{\rho C_P} \frac{\partial^2 T}{\partial x^2} \qquad \text{(B-10)}$$

where,

T = temperature in °F
t = time in hours
k = thermal conductivity in Btu/(hr-ft-°F)
ρ = density in Lb_m/ft^3
C_p = specific heat in Btu/(lb_m-°F)
x = linear dimension in ft

also,

$$\alpha = \frac{k}{\rho C_p} = \text{thermal diffusivity}$$

Therefore equation (B-10) can be re-written to read:

$$\frac{\partial T}{\partial t} = \alpha \frac{\partial^2 T}{\partial x^2} \qquad (B\text{-}11)$$

Equation (B-11) is used to solve the heat transfer problem shown in Figure B-3.

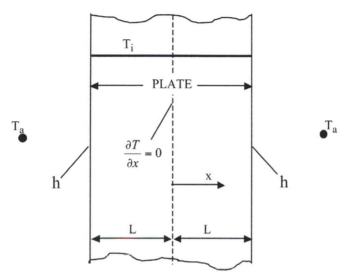

Figure B-3. Plate initially at uniform temperature T_i suddenly exposed to convective environment at constant temperature, T_a

The temperature response of the plate shown in Figure B-3 is symmetric about the center line, therefore only one half of the plate needs to be analyzed. The solution of this problem is given by Chapman[1], as follows:

$$\frac{T - T_a}{T_i - T_a} = 2 \sum_{n=1}^{\infty} \frac{e^{-\delta_n^2 \alpha t L^{-2}} \sin \delta_n \cos\left(\delta_n \frac{x}{L}\right)}{\delta_n + \sin \delta_n \cos \delta_n} \qquad (B\text{-}12)$$

In equation (B-12) T is the plate temperature as a function of position and time, T_i is the initial plate temperature, T_a is the temperature of the fluid surrounding the plate, t is time, L is one half of the plate thickness, and x is the distance from the plate center line. The term α is the plate's thermal diffusivity, defined above, and δ_n is defined by the following equation:

$$\delta_n \tan \delta_n = hL/k \qquad\qquad (B\text{-}13)$$

where,

 h = the convective film coefficient, Btu/(hr-ft^2-°F)
 k = thermal conductivity in Btu/(hr-ft-°F)
 L = one half of the plate thickness, ft

An example problem of the plate shown in Figure B-3 follows. Assume the following values.

Table B-2. Material properties, temperatures, and
dimensions for example problem

THICKNESS (L), in	2
DENSITY, Lb$_m$/ft^3	484
CONDUCTIVITY, Btu/(hr-ft-°F)	21
SP. HEAT, Btu/(lb$_m$-°F)	0.116
FILM COEF., Btu/(hr-ft^2-°F)	20
INTIAL SLAB TEMP., °F	600
FLUID ENVIRONMENT TEMP., °F	100
x, in	1

Solving equation (B-13) the following values are established for Δ_n.

Table B-3. Values of Δ_n derived from equation (B-13)

n	δ_n
1	3.19129
2	6.308342
3	9.441588
4	12.57899
5	15.71806
6	18.85797
7	21.99836

Using equation (B-12), the following results are calculated:

Table B-4. Analytical solution of the plate problem

TIME, sec	TEMP., °F
0	600
20	597.23
40	591.75
60	586.21
80	580.74
100	575.34
120	570.01
140	564.74
160	559.53
180	554.37

The solution of the problem given above is difficult, particularly the evaluation of Δ_n. Many heat transfer problems similar to one given above have been solved and their solutions are shown in reference 2.

NUMERICAL SOLUTION— TRANSIENT WITH CONDUCTION AND CONVECTION

A numerical solution of the plate problem shown in Figure B-3 is developed by breaking the plate (half plate) into small pieces, or nodes. The numerical equivalent of equation (B-10) is then applied to each of the nodes. Equation (B-10) is re-written:

$$\rho C_P \frac{\partial T}{\partial t} = k \frac{\partial^2 T}{\partial x^2} \tag{B-14}$$

In equation (B-14) T is temperature in °F, t is time in hours, x is distance in ft, ρ is density in lb_m/ft^3, C_P is specific heat in $Btu/(lb_m\text{-}°F)$, and k is thermal conductivity in $Btu/(hr\text{-}ft\text{-}°F)$. Working with the partial derivatives:

$$\frac{\partial T}{\partial t} \approx \frac{T'_x - T_x}{\Delta t} \tag{B-15}$$

Δt is a small but finite step in time.

$$\frac{\partial^2 T}{\partial x^2} \approx \frac{\left(\frac{\Delta T}{\Delta x}\right)_{x+1} - \left(\frac{\Delta T}{\Delta x}\right)_x}{\Delta x}$$

Δx is a small but finite step in distance.

$$\frac{\partial^2 T}{\partial x^2} \approx \frac{\frac{T_{x+1} - T_x}{\Delta x} - \frac{T_x - T_{x-1}}{\Delta x}}{\Delta x} \tag{B-16}$$

$$\frac{\partial^2 T}{\partial x^2} \approx \frac{1}{\Delta x^2}\left[(T_{x+1} - T_x) + (T_{x-1} - T_x)\right]$$

Equations (B-15) and (B-16) are substituted into equation (B-14).

$$\rho C_P \frac{T'_x - T_x}{\Delta t} \approx \frac{k}{\Delta x^2}\left[(T_{x+1} - T_x) + (T_{x-1} - T_x)\right] \tag{B-17}$$

Multiply both sides of equation (B-17) by $A\Delta x$, yields:

$$\rho A \Delta x C_P \frac{T'_x - T_x}{\Delta t} \approx \frac{kA}{\Delta x}\left[(T_{x+1} - T_x) + (T_{x-1} - T_x)\right] \tag{B-18}$$

This previous equation reduces to:

$$M C_P \frac{T'_x - T_x}{\Delta t} \approx \frac{kA}{\Delta x}(T_{x+1} - T_x) + \frac{kA}{\Delta x}(T_{x-1} - T_x) \tag{B-19}$$

M = the mass of the node = $\rho A \Delta x$ = density time volume

At this point there is a decision to make relative to the temperatures on the right side of equation (B-19). On the left side of the equation there are the temperatures T_x' and T_x. T_x' is the temperature of node x at the new time after the time step, Δt. T_x is the temperature of node x initially, or at the old time. The temperatures on the right side of equation (B-19) can be assigned to the old time or the new time or a combination of the old and new times. If the temperatures on the right side of equation (B-19) are taken to be at the old time we have the forward differencing or explicit approach. Taking this approach we can solve for T'_x in terms of temperatures all at the old time, as shown in equation (B-20).

$$T'_x = T_x + \frac{\Delta t}{MC_p} \left[\frac{kA}{\Delta x} \left(T_{x+1} - T_x \right) + \frac{kA}{\Delta x} \left(T_{x-1} - T_x \right) \right] \qquad \text{(B-20)}$$

Similar equations can be written for each node point and the problem solved by repeatedly solving the equations. It is assumed that the initial temperatures at each node point are known and the solution of the defining equations (like equation (B-20) yields the temperatures at time, Δt. This process is repeated to yield the temperatures at $2\Delta t$, $3\Delta t$, etc.

If instead of assuming the temperatures on the right side of equation (B-20) are at the old time it is assumed they are at the new time, we have equation (B-21).

$$T'_x = T_x + \frac{\Delta t}{MC_p} \left[\frac{kA}{\Delta x} \left(T'_{x+1} - T'_x \right) + \frac{kA}{\Delta x} \left(T'_{x-1} - T'_x \right) \right]$$

$$\text{(B-21)}$$

Equation (B-21) represents the backward differencing or implicit approach. Equations similar to equation (B-21) can be written for each of the nodes resulting in n equations for n nodes. This set of equations must be solved simultaneously to obtain the temperatures at the new time, Δt. After solving for the temperatures at the new time the process is repeated to find the temperatures at $2\Delta t$, $3\Delta t$, and on to the maximum time. There are advantages and disadvantages to the forward and backward differencing approaches. These will be discussed later.

Returning now to equation (B-20) and rearranging we have:

$$T'_x = \frac{\Delta t}{MC_p}\left[\frac{kA}{\Delta x}\left(T_{x+1}+T_{x-1}\right)\right]+T_x\left(1-\frac{2kA/\Delta x}{MC_p}\Delta t\right) \qquad \text{(B-22)}$$

Equation (B-19) suggests the thermal network given below. The $kA/\Delta x$ quantities are thermal conductors and the item labeled MC_P is a thermal capacitor.

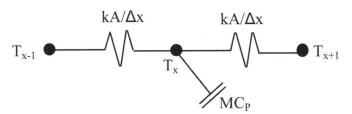

Figure B-4. Thermal network representing equation (19)

There is a restriction on the size of the time step which can be employed when using equation (B-22), the forward differencing approach. Consider the last term to the right in equation (B-22):

$$\left(1-\frac{2kA/\Delta x}{MC_P}\Delta t\right)$$

If this term becomes negative, the solution for the temperature at x at the new time, T'_x, may diverge and give incorrect values. This can be avoided by choosing a value of Δt that guarantees that the term is always positive. Therefore:

$$\Delta t \leq \frac{MC_P}{2kA/\Delta x} \tag{B-23}$$

A more general version of the equation given above is:

$$\Delta t_{crit} \leq \frac{MC_P}{\sum\limits_{1}^{N} G_N} \tag{B-24}$$

In equation (B-24) Δt_{crit} is the critical time step, M the mass of the node, C_P the specific heat of the node material, G_N the thermal conductances between the node being evaluated and its neighboring nodes. Thermal conductances can be of the form $kA/\Delta x$ for thermal conduction or hA for thermal convection. The compute time step used when applying the forward differencing approach must be equal to are less than the critical time step and it is considered good practice to use a compute time step equal to are less that one half the critical time step.

The time step constraint for the forward differencing approach does not apply to the backward differencing approach. However, it is good practice, when using the backward differencing approach to not take time steps greater than 5 or 10 times the critical time step defined by equation (B-24). Also, the backward differencing approach requires the solution of sets of simultaneous equations while the forward differencing approach uses a relatively simple algebraic equation at each node for each time step. It is the author's opinion that the forward differencing approach is generally the best approach when using Microsoft Excel to solve heat transfer problems.

The problem defined in Figure B-3 will be solved numerically using both the forward and backward differencing approaches and compared to the analytical solution. The thermal network given below defines the numerical solution.

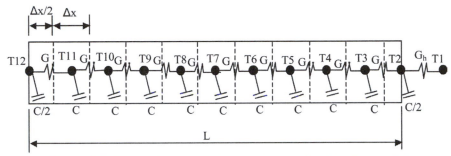

Figure B-5. Thermal network for problem defined in Figure B-3

In Figure B-5, $G = kA/\Delta x$ (thermal conductance-conduction), $G_h = hA$ (thermal conductance-convection), $C = MC_P$ (thermal capacitance), and Δx is the node spacing. The values defining the problem are listed in Table B-2 and the numerical solution is given in Table B-5.

Table B-5. Numerical and analytical solutions of plate problem

Time, sec	FORWARD DIFF. SOLUTION $\Delta t = 0.5$ sec °F	BACKWARD DIFF. SOLUTION $\Delta t = 5.0$ sec °F	ANALY SOLUTION °F
0	600	600	600
20	597.3	597.1	579.2
40	591.8	591.8	591.8
60	586.3	586.3	586.2
80	580.8	580.7	580.7
100	575.4	575.3	575.3
120	570.1	570.0	570.0
140	564.8	564.7	564.7
160	559.6	559.5	559.5
180	554.4	554.3	554.4

It is evident from the results shown in Table B-5 that the numerical solutions are very good approximations of the exact analytical solution. The critical time step for the plate problem is

0.668 seconds. Even though the time steps taken using the backward differencing approach were 10 times the time steps taken using the forward differencing approach, it took less time to set up and solve the problem on Microsoft Excel using the forward differencing approach.

ANALYTICAL—
STEADY STATE FLUID FLOW IN A TUBE

Next we will derive an equation for the fluid temperature as a function of the distance along the tube. Consider an energy balance at the fluid element shown in Figure B-6.

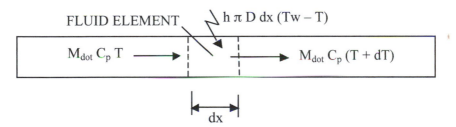

Figure B-6. Energy balance for flow in a tube

An energy balance on the fluid element results in the following equation:

$$M_{dot} \, C_p \, T - M_{dot} \, C_p \, (T + dT) + h \, \pi \, D \, dx \, (T_w - T) = 0 \quad \text{(B-25)}$$

Where,

M_{dot} = mass rate of flow of the fluid = volume rate of flow times fluid density
C_p = specific heat of the fluid
T = fluid temperature
dT = differential temperature change
h = film coefficient
D = tube inside diameter

dx   = differential length of tube
π D dx = surface area tube at fluid element
T$_w$ = tube wall temperature (a constant)

The energy balance equation reduces to:

$$- M_{dot} \, C_p \, dT = - h \, \pi \, D \, dx \, (T_w - T)$$

Let $T_w - T = T_b$, then $dT_b = - dT$

Substituting:

$$M_{dot} \, C_p \, dT_b = - h \, \pi \, D \, dx \, T_b$$

Rearranging:

$$dT_b / T_b = - (h \, \pi \, D / M_{dot} \, C_p) \, dx$$

Integrating:

$$\ln T_b = - (h \, \pi \, D / M_{dot} \, C_p) \, x + \text{constant, let the constant} =$$
$$\ln C$$

$$\ln (T_b / C) = - (h \, \pi \, D \, x / M_{dot} \, C_p)$$

$$Tb / C = e^{(- h \, \pi \, D \, x / Mdot \, Cp)}$$

$$Tw - T = C \, e^{(- h \, \pi \, D \, x / Mdot \, Cp)}$$

At $x = 0$, $T = T_i$; T_i = the initial temperature. This results in the equation for flow in a tube with a constant wall temperature and a constant film coefficient.

$$T = T_w + (T_i - T_w) \, e^{(- h \, \pi \, D \, x / Mdot \, Cp)} \qquad \text{(B-26)}$$

NUMERICAL—
STEADY STATE FLUID FLOW IN A TUBE

We will now derive an equation for the fluid temperature as a function of the distance along the tube, just as was done in the previous section. Consider an energy balance at the fluid element shown in Figure B-7.

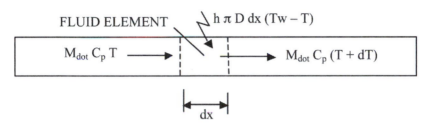

Figure B-7. Energy balance for flow in a tube.

An energy balance on the fluid element results in the following equation:

$$M_{dot} \, C_p \, T - M_{dot} \, C_p \, (T + dT) + h \, \pi \, D \, dx \, (T_w - T) = 0$$

Where,

M_{dot} = mass rate of flow of the fluid = volume rate of flow times fluid density

C_p = specific heat of the fluid

T = fluid temperature

dT = differential temperature change

h = film coefficient

D = tube inside diameter

dx = differential length of tube

$\pi \, D \, dx$ = surface area tube at fluid element

T_w = tube wall temperature (a constant)

The energy balance equation reduces to:

$$- M_{dot} \, C_p \, dT = - h \, \pi \, D \, dx \, (T_w - T)$$

We will approximate the infinitesimal equation, above, with a finite difference equation by letting $dT \approx T_{x+\Delta x} - T_x$ and $dx \approx \Delta x$. Rearranging the terms gives:

$$M_{dot}\, C_p\, (T_x - T_{x+\Delta x}) + h\, \pi\, D\, \Delta x (T_w - T_{x+\Delta x}) = 0 \qquad \text{(B-27)}$$

Equation (B-27) can be represented by the thermal network shown in Figure B-8.

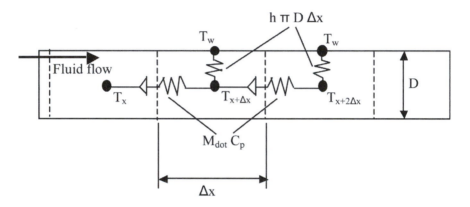

Figure B-8. Thermal network for fluid flow in a tube

The $M_{dot}\, C_p$ conductors shown in Figure B-8 are one way conductors. This is consistent with Eq. (B-27) which shows that node $x + \Delta x$ is influenced only by node T_x and node T_w but not by node $T_{x+2\Delta x}$. Note that T_x is an upstream node relative to node $T_{x+\Delta x}$ while node $T_{x+2\Delta x}$ is downstream to node $T_{x+\Delta x}$.

We will now solve a sample problem numerically and compare the solution with the analytical solution. The thermal network which characterizes the problem is given in Figure B-9 and the quantities that define the problem are given is Table B-6. A comparison of the numerical and analytical solutions is shown in Table B-7. The analytical solutions are from equation (B-26). The numerical solution is fairly close to the analytical solution and can be made more accurate by increasing the number of nodes.

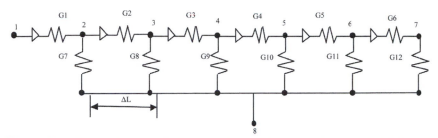

Figure B-9. Thermal network for sample problem, steady-state fluid flow in a tube

Table B-6. Material properties, temperatures, and dimensions for the steady state, fluid flow example problem

DENSITY, LBm/CU FT	61.99
VISCOSITY, LBm/(FT-HR)	1.6500
CONDUCTIVITY,BTU/(HR-FT-°F)	0.3640
SPECIFIC HEAT, BTU/(LBm-°F)	0.997
PRANDTL NUMBER	4.519
TUBE DIA, IN	1
TUBE DIA, FT	0.0833333
FLUID VEL., FT/SEC	1
FILM COEF., BTU/HR-FT^2-F	100
TUBE LENGTH, IN	120
TUBE LENGTH, FT	10
DELTA L, IN	20
DELTA L, FT	1.6666667
MASS RATE OF FLOW, LB/SEC	0.338103
REYNOLDS NUMBER	11270.909

Table B-7. Numerical and analytical solution of the steady state, fluid flow problem

	NUM SOLUTION °F	ANALY SOLUTION °F	% DIFF	DISTANCE FROM INLET, IN
T1	100	100	0	0
T2	103.4708	103.53172	1.72%	20
T3	106.8211	106.93871	1.69%	40
T4	110.0552	110.22538	1.66%	60
T5	113.177	113.39596	1.63%	80
T6	116.1904	116.45458	1.61%	100
T7	119.0993	119.40517	1.58%	120
T8	200	200	(AMBIENT TEMP)	

ANALYTICAL—
TRANSIENT FLUID FLOW IN A TUBE

Consider a tube, a fluid flowing in the tube, and the surrounding environment all at an initial temperature of zero, shown
in Figure B-10. At time equal to zero the fluid inlet temperature
steps up to a constant value, T_o. The thermal physical properties of the tube and fluid are assumed to be constant as well as
the fluid flow rate. There is convection from the flowing fluid
to the tube wall and from the tube wall to the surrounding environment. The film coefficients from the fluid to the tube and
from the tube to the surrounding environment are assumed to
be constant. Neglect axial thermal conduction in the fluid and
the tube. Assume that at any x location the fluid temperature
is represented by a bulk (or average) fluid temperature and the
radial temperature change across the tube is zero. The surrounding environment temperature is constant and equal to zero. The
analytical solution for the problem described above is given in
reference 3, starting on page 353.

When $t < x/V$,

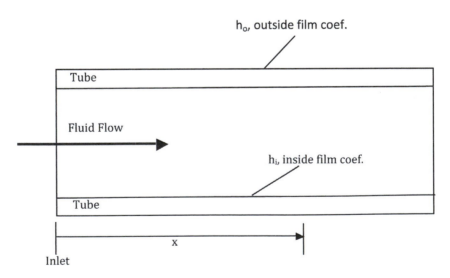

Figure B-10. Sample problem for transient flow in a tube

$$\theta(x,t) = T(x,t) = 0$$

$\theta(x,t)$ is the tube temperature as a function of distance from inlet (x), and time (t)

$T(x,t)$ is the fluid temperature as a function of distance from inlet (x), and time (t)

V is the average fluid velocity in the tube, a constant

x/V is the time it takes the fluid to flow the distance x

When $t > x/V$,

$$\frac{\theta(x,t)}{T_o} = b_2 e^{-b_1 x/V} \int_0^{t^*} e^{-b_4 s} I_0 \left[2\left(\frac{b_1 b_2 xs}{V}\right)^{\frac{1}{2}} \right] ds \tag{B-28}$$

$$\frac{T(x,t)}{T_o} = e^{-b_1 x/V} \left\{ e^{-b_4 t^*} I_0 \left[2\left(\frac{b_1 b_2 xt^*}{V}\right)^{\frac{1}{2}} \right] + b_4 \int_0^{t^*} e^{-b_4 s} I_0 \left[2\left(\frac{b_1 b_2 xs}{V}\right)^{\frac{1}{2}} \right] ds \right\} \tag{B-29}$$

$b_1 = h_i P_i / \rho_f c_f A_f$
$b_2 = h_i P_i / \rho_t c_t A_t$
$b_3 = h_o P_o / \rho_t c_t A_t$
$b_4 = b_2 + b_3$

$t^* = t - x/V$
h_i = the film coefficient between the fluid inside the tube and the tube inner wall, Btu/(hr-ft2-°F)
h_o = the film coefficient between the fluid outside the tube and the tube outer wall, Btu/(hr-ft2-°F)
P_i = inside perimeter of the tube, ft
P_o = outside perimeter of the tube, ft
ρ_f = density of fluid flowing inside the tube, Lb$_m$/ft3
c_f = specific heat of fluid flowing inside the tube, Btu/(lb$_m$-°F)

A_f = cross sectional area of tube interior, ft^2
ρ_t = density the tube material, Lb$_m$/ft^3
c_t = specific heat of the tube material, Btu/(lb$_m$-°F)

A_t = cross sectional area of tube, ft^2
s = variable of integration, sec
I_o = modified Bessel function
T_o = temperature the fluid is stepped to at time zero, °F

The quantities that define the transient, fluid flow example problem are given in Table B-8. The analytical solution for the tube temperature at 30 feet from the inlet is shown in Figure B-11. Computing the values of the tube temperature as a function of time using Eq. (B-28) is not an easy task. This calculation requires the numerical integration of a function containing a modified Bessel Function. For the values developed here 200 divisions were used in the integration.

Table B-8. Material properties, temperatures, and dimensions for the transient, fluid flow example problem

TUBE ID, FT	0.067583
TUBE OD, FT	0.072917
FLUID VEL, FT/SEC	0.692497
INSIDE FILM COEF, BTU/(HR-FT^2-F)	263.3
OUTSIDE FILM COEF, BTU/(HR-FT^2-F)	1.4
TUBE DENSITY, LB/FT^3	559
TUBE SP. HEAT, BTU/(LB-F)	0.091
TUBE CROSS SECTIONAL AREA, FT^2	0.000589
OUTSIDE TUBE PERIMETER, FT	0.229074
FLUID DENSITY, LB/FT^3	61.4
FLUID SP. HEAT, BTU/(LB-F)	1
FLUID CROSS SECTIONAL AREA, FT^2	0.003587
INSIDE TUBE PERIMETER, FT	0.212319
To, INLET TEMP., F	100
INITIAL TUBE AND FLUID TEMP. = 0	0
DISTANCE, FT	30
t*, sec	11.67851

NUMERICAL—
TRANSIENT FLUID FLOW IN A TUBE

The thermal network used in conjunction with the numerical solution of the transient, fluid flow example problem is shown in Figure B-11. The values of the thermal conductors, thermal capacitors, and other items associated with the thermal network are given in Table B-9.

In Figure B-12 two numerical solutions and the analytical solution for the transient, fluid flow example problem are shown. The numerical solutions do not match the analytical solution as well as we have seen for the other comparisons shown earlier in this text. The forward differencing approach was used to numerically solve the transient fluid flow sample problem. Inspection of Figure B-12 shows that the numerical solution using the critical time step as the compute time step is more accurate that when ½ the critical time step is used as the compute time step. To understand why the difference exists between the analytical and numerical solutions and between the numerical solutions we need to compute the critical time step, Δt_{crit}, for the fluid nodes.

$$\Delta t_{crit} \leq \frac{C}{\Sigma G} \qquad \text{(B-30)}$$

The quantity C is the thermal capacitance of the given node and the G's are the thermal conductors connected to the node. For the fluid nodes in Figure B-11,

$$\Delta t_{crit} \leq \frac{M\,c_p}{(M_{dot}\,c_p + h_i\,A_i)} \qquad \text{(B-31)}$$

M = mass of fluid node
c_p = specific heat of the fluid
M_{dot} = mass rate of flow of the fluid
h_i = inside film coefficient (between fluid and inside tube wall)

A_i = inside surface area of tube for the given node

Equation (B-31) is rewritten, noting the following:

$M = \rho \, \Delta x \, A_t$

$M_{dot} = \rho \, V \, A_t$

Where:

ρ = fluid density

Δx = length of fluid node

A_t = cross sectional area of inside of pipe

V = fluid velocity

$$\Delta t_{crit} = \frac{\rho A_t \Delta x c_p}{\rho V A_t c_p + h_i A_i} = \frac{\Delta x / V}{1 + \dfrac{h_i A_t}{\rho V A_c c_p}} \tag{B-32}$$

The quantity $\Delta x / V$ is the time it takes the fluid to flow the length of the node. Since $h_i A_t / \rho V A_c c_p$ is a positive number, it follows that the critical time step must be smaller than the time it takes the fluid to flow from node to node. Therefore, thermal effects at some given point in the system are propagated downstream by the forward differencing analysis faster than the fluid is flowing. For example, in Figure B-12 the predictions by the forward differencing approach show an effect at the given node before the fluid has had time to reach the node. To further explain, assume that the critical time step is one half the time it takes the fluid to flow from node to node. Now in Figure B-11 let the fluid node T_1 step in temperature at time equal zero. At the next compute time step T_2 will feel an effect of the step change at T_1 even though the fluid has not had time to reach T_2. This effect continues as time advances and the adverse effect compounds.

It has been shown that the problem with forward differencing is that one is forced to use a compute time step smaller than the time it takes the fluid to flow from node to node. If the computer time step is taken to be ½ the critical time step (for most problems this is considered good practice) inaccuracies are greater than if the critical time step is used for the compute time step. This effect is shown in Figure B-12. If the forward differencing approach produces inherent errors why not then use the backward differencing scheme that does not have a critical time step requirement? The backward differencing numerical approach

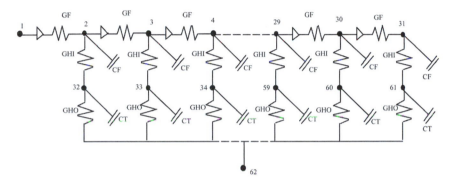

Figure B-11. Thermal network for sample problem, transient fluid flow in a tube

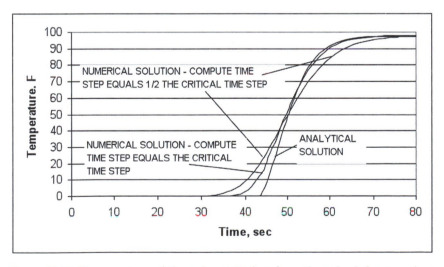

Figure B-12. Temperature of the tube at 30 feet from the tube inlet, transient, fluid flow example problem

allows the use of a time step equal to the time it takes the fluid to flow from node to node. The problem with the backward differencing technique is that all the nodes are computed simultaneously at each time step. This means that the effect, say, of a step change in the fluid temperature at the inlet is felt downstream throughout the fluid system at the very next time step.

Table B-9. Numerical values for the quantities shown in the thermal network, Figure B-10.

NODE LENGTH, ft	1
GF, FLOW CONDUCTOR, Btu/(sec-°F)	0.15253
(MASS RATE OF FLOW TIMES SPECIFIC HEAT)	
GHI, INSIDE FILM COEFFICIENT CONDUCTOR, Btu/(sec-°F)	0.015529
(INSIDE FILM COEF. TIMES NODE SURFACE AREA)	
GHO, OUTSIDE FILM COEFFICIENT CONDUCTOR, Btu/(sec-°F)	8.91E-05
(OUTSIDE FILM COEF. TIMES THE NODE SURFACE AREA)	
CF, FLUID THERMAL CAPACITANCE, Btu/°F	0.22026
(MASS OF FLUID NODE TIMES FLUID SPECIFIC HEAT)	
CT, TUBE THERMAL CAPACITANCE, Btu/°F	0.029938
(MASS OF TUBE NODE TIMES TUBE SPECIFIC HEAT)	
CRIT TIME STEPS, sec	
FLUID	1.311
TUBE	1.917
TIME FOR FLUID TO TRAVEL THE LENGTH OF A NODE, sec	1.444
TIME STEP, sec	0.6553

RADIATION NETWORKS

The characteristics of radiation (Oppenheim[4]) networks are such that at each given surface, say surface x, the adjacent conductor is of the form $A_x\varepsilon_x/(1 - \varepsilon_x)$, where A is area and ε is emissivity of the surface. This conductor is connected between the surface node and the dummy J_x node. The potential at the surface node is $\sigma(T_x)^4$, where σ is the Stefan-Boltzmann constant and T is the surface absolute temperature. The network is completed by connecting each J node to every other J node. The conductors connecting the J nodes are of the form $A_x F_{x-y}$, where A is area and F the view factor. If a given surface is perfectly insulated or if the emissivity of the node is one (a black surface), the conductor, $A_x\varepsilon_x/(1 - \varepsilon_x)$, is eliminated. The effect is, therefore, that the surface potential, $\sigma(T_x)^4$, moves to the J node location.

A radiation network is shown in Figure B-13. The three surfaces, x, y, and z shown in the figure are floating in space. Surfaces x and y are held at constant temperatures with heaters and thermostats. Surface z is perfectly insulated. The node labeled σT_s^4 represents deep space with a temperature of approximately zeros degrees absolute and an emissivity of 1. For surface z, the perfectly insulated surface, there can be no heat flow to or from the surface. Since the value of the conductor, $A_z\varepsilon_z/(1 - \varepsilon_z)$, is a finite number, then σT_z^4 must equal J_z to insure no heat transfer to or from surface z. This eliminates the $A_z\varepsilon_z/(1 - \varepsilon_z)$ conductor and σT_z^4 moves to the J_z node. The value of the conductor, $A_s\varepsilon_s/(1 - \varepsilon_s)$, is infinity since the emissivity of space, ε_s, is equal to one. This "shorts out" the conductor and the space node moves to the J_s node point. The radiation network shown in Figure B-13 can be simplified and the modified network is shown in Figure B-14.

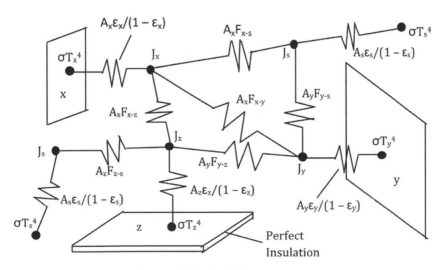

Figure B-13. Radiation Network

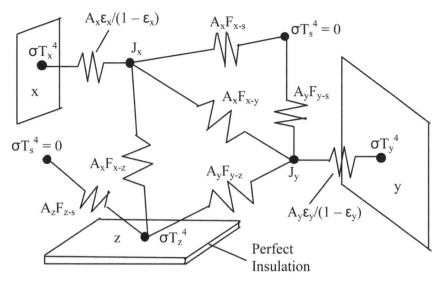

Figure B-14. Simplified Radiation Network

The following values are assumed for the temperatures, radiation view factors, areas, and emissivities shown in Figure B-14.

$T_x = 720°R$	$\sigma T_x^4 = 4.606E02$
$T_y = 1440°R$	$\sigma T_y^4 = 7.37E3$
$A_x = 21.53$ ft^2	$A_x\varepsilon_x/(1 - \varepsilon_x) = 86.12$
$A_y = 43.06$ ft^2	$A_y\varepsilon_y/(1 - \varepsilon_y) = 387.54$
$A_z = 86.11$ ft^2	$A_xF_{x-y} = 4.306$
$\varepsilon_x = 0.8$	$A_xF_{x-z} = 6.459$
$\varepsilon_y = 0.9$	$A_xF_{x-s} = 10.77$
$F_{x-y} = 0.2$	$A_yF_{y-z} = 4.306$
$F_{x-z} = 0.3$	$A_yF_{y-s} = 34.45$
$F_{x-s} = 0.5$	$A_zF_{z-s} = 75.35$
$F_{y-z} = 0.1$	
$F_{y-s} = 0.8$	
$F_{z-s} = 0.875$	

We now write a steady state energy balance at each node whose potential (temperature) is unknown. Note, the nodes labeled $\sigma T_x{}^4$ and $\sigma T_y{}^4$ are known since the temperatures of these nodes are given. For node J_x we have:

$$A_x\varepsilon_x/(1 - \varepsilon_x)(\sigma T_x{}^4 - J_x) + A_xF_{x\text{-}s}(0 - J_x) + A_xF_{x\text{-}y}(J_y - J_x) + A_xF_{x\text{-}z}(\sigma T_z{}^4 - J_x) = 0$$

Solving for J_x, we have:

$$J_x = [A_x\varepsilon_x/(1 - \varepsilon_x)\sigma T_x{}^4 + A_xF_{x\text{-}y}J_y + A_xF_{x\text{-}z}\sigma T_z{}^4]/[A_x\varepsilon_x/(1 - \varepsilon_x) + A_xF_{x\text{-}s} + A_xF_{x\text{-}y} + A_xF_{x\text{-}z}]$$

$$J_x = [86.12{*}460.6 + 4.306{*}Jy + 6.459{*}\ \sigma T_z{}^4]/[86.12 + 10.77 + 4.306 + 6.459] = [39666.9 + 4.306{*}Jy + 6.459{*}\ \sigma T_z{}^4]/107.655 \tag{B-33}$$

Energy balances are written at node J_y and node $\sigma T_z{}^4$. The resulting equations are solved for J_y and $\sigma T_z{}^4$ to yield:

$$J_y = [2.856\text{E}06 + 4.306J_x + 4.306\ \sigma T_z{}^4]/430.6 \tag{B-34}$$

$$\sigma T_z{}^4 = [6.459{*}J_x + 4.306{*}J_y]/86.115 \tag{B-35}$$

It is assumed that the values of the node points are 10000 as an initial guess. Equations (B-33), (B-34), and (B-35) are solved by the relaxation method using the initial guessed values to yield updated values. The relaxation process is repeated until the calculated values converge. For the problem at hand it took eight iterations to converge to the following results:

$$J_x = 657.06$$
$$J_y = 6643.0$$
$$\sigma T_z{}^4 = 381.45 \qquad\qquad T_z = 686.8{°}\text{R}$$

Another way to approach the radiation problem is to solve the three equations (B-33), (B-34), and (B-35) simultaneously. Rewriting the equations we have:

$$107.655Jx - 4.306Jy - 6.459^* \, \sigma T_z{}^4 = 39666.9 \qquad\qquad \text{(B-36)}$$

$$- \, 4.306Jx + 430.6Jy - 4.306^*\sigma T_z{}^4 = 2.856E06 \qquad\qquad \text{(B-37)}$$

$$- \, 6.459^*J_x - 4.306^*J_y + 86.115^* \, \sigma T_z{}^4 = 0 \qquad\qquad \text{(B-38)}$$

Simultaneous equations can be solved easily using Microsoft Excel. First invert the matrix defined by the left sides of the set of equations using the Excel function MINVERSE. The matrix representing the left side of equations (B-36), (B-37), and (B-38) is as follows:

107.655 – 4.306 – 6.459
– 4.306 + 430.6 – 4.306
– 6.459 – 4.306 + 86.115

Inverting the above matrix yields:

0.009335	0.000100	0.000705
0.000100	0.002325	0.000124
0.000705	0.000124	0.011671

The inverted matrix is then multiplied by the column matrix defined by the right hand sides of equations (B-36), (B-37), and (B-38).

39666.9
2.86E06
0

The matrix multiplication is accomplished using the Excel function, MMULT. The results of the simultaneous solution are exactly the same as the relaxation solution. The Microsoft Excel program used for the matrix manipulations is from Office xp, version 2002. This version of Excel can handle matrices up to a size of 52x52.

NUMERICAL—COMBINING CONDUCTION, CONVECTION, AND RADIATION NETWORKS (STEADY STATE)

Heat transfer problems can involve conduction, convection, and radiation modes of heat transfer. This forces the coupling of conduction, convection, and radiation and makes it highly desirable to combine the conduction, convection, and radiation networks. The problem with combining the networks is that the potentials are vastly different; conduction and convection to the first power and radiation to the fourth power of the temperature. This difference is overcome by manipulating the radiation relationship. In general the radiation equation is given by q = (a coefficient) times $(\sigma T_x^4 - \sigma T_y^4)$. The coefficient is generally of the form $\varepsilon_x A_x/(1 - \varepsilon_x)$ or $A_x F_{x-y}$. Working with the temperature to the fourth power term, first the Stefan-Boltzmann constant is factored and then the remaining temperature term is also factored to yield: $\sigma[(T_x^2 + T_y^2)(T_x + T_y)](T_x - T_y)$. This operation transforms the radiation equation to a form that includes the temperature difference to the first power with the remaining terms forming a temperature-dependent conductor. The radiation conductor values are given by:

$$\left[\frac{\sigma \varepsilon_x A_x}{1 - \varepsilon_x}\right]\left[\left(T_x^2 + T_y^2\right)\left(T_x + T_y\right)\right]$$

or

$$\left[\sigma A_x F_{x-y}\right]\left[\left(T_x^2 + T_y^2\right)\left(T_x + T_y\right)\right]$$

The manipulation of radiation expressions to include the temperature difference to the first power allows the combination of conduction, convection, and radiation thermal networks and aids in the solution of complex problems.

An example problem which includes conduction, convection, and radiation follows. Consider the cylindrical fin shown in Figure B-15. The fin is 0.0164 ft in diameter, 0.3937 ft long and made of alloy 1100 aluminum with a thermal conductivity of 128.3 Btu/(hr-ft-°F).

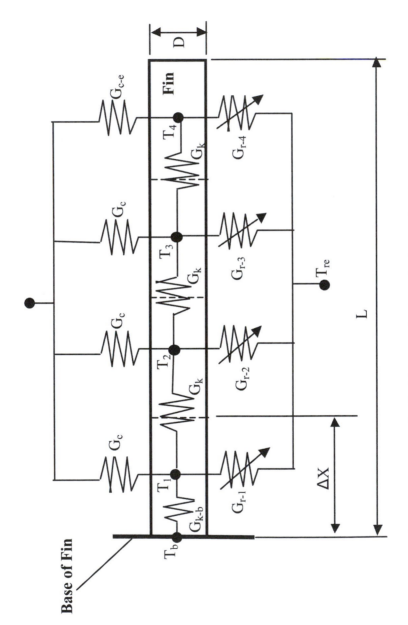

Figure 15. Heat transfer fin with combined conduction, convection, and radiation

The node length, ΔX, is 0.09843 ft. The temperature of the base of the fin is held constant at 440.3°F, the convective environment temperature and the effective radiation environmental temperature are the same at 80.31°F. It is assumed that the view factors, $F_{1\text{-}e}$, $F_{2\text{-}e}$, $F_{3\text{-}e}$, and $F_{4\text{-}e}$ are all equal to one. Also, it is assumed that the aluminum fin is painted with a black paint with an emissivity of 0.95.

The quantities shown in Figure B-15 are defined as follows:

T_b = the temperature of the base of the fin = 440.3°F

T_1, T_2, T_3, and T_4 are the temperatures along the fin

G_k = the conduction conductor along the fin = $kA/\Delta X$ = 7.635E – 05 Btu/(sec-°F)

k = thermal conductivity of the fin material = 128.3 Btu/(hr-ft-°F)

A = cross sectional area of the fin = $\pi D^2/4$ = 2.113E – 04 ft^2

D = diameter of the fin = 1.64E – 02 ft

ΔX = length of the fin's node = 9.843E – 02 ft

$G_{k\text{-}b}$ = the conduction conductor at the base of the fin = 1.53E – 04 Btu/(sec-°F)

G_c = convection conductor along the fin = hA_s =3.7221E – 06 Btu/(sec-°F)

h = film coefficient, 2.642 Btu/(hr-ft^2-°F)

A_s = surface area of the node = $\pi D \Delta x$ = 4.7124E – 04 m^2

$Gc - e$ = convection conductor at end of the fin = $h(A_s + A)$ = 3.8772E – 06 Btu/(sec-°F)

A = surface area of end of the fin = cross sectional area of the fin

$G_{r\text{-}1}$ = radiation conductor = $\sigma A_s \varepsilon F_{1\text{-}e}(T_1^2 + T_{re}^2)(T_1 + T_{re})$ Btu/(sec-°F)

$G_{r\text{-}2}$ = radiation conductor = $\sigma A_s \varepsilon F_{2\text{-}e}(T_2^2 + T_{re}^2)(T_2 + T_{re})$ Btu/(sec-°F)

$G_{r\text{-}3}$ = radiation conductor = $\sigma A_s \varepsilon F_{3\text{-}e}(T_3^2 + T_{re}^2)(T_3 + T_{re})$ Btu/(sec-°F)

$$G_{r-4} = \text{radiation conductor} = \sigma(As + A)\varepsilon F_{4-e}(T_4^2 + T_{re}^2)$$
$$(T_4 + T_{re}) \text{ Btu}/(\text{sec-}°\text{F})$$
$$\sigma A_s \varepsilon F_{1-e} = \sigma A_s \varepsilon F_{2-e} = \sigma A_s \varepsilon F_{3-e} = 2.294\text{E} - 15 \text{ Btu}/(\text{sec-R}^4)$$
$$\sigma(As + A)\varepsilon F_{4-e} = 2.3897\text{E} - 15 \text{ Btu}/(\text{sec-R}^4)$$

σ = Stefan-Boltzmann constant = $0.1714\text{E} - 08$ Btu/ (hr-ft2-°R4)

ε = emissivity of fin surface = 0.95

T_{re} = effective temperature of the radiation environ- ment = $80.31°\text{F}$

T_{ce} = temperature of the convective environment = $80.31°\text{F}$

The radiation conductor shown above is derived by using the Oppenheim Radiation Network shown in Figure B-16.

σT_1^4 $A_1\varepsilon_1/(1-\varepsilon_1)$ A_1F_{1-r} $A_r\varepsilon_r/(1-\varepsilon_r)$ σT_r^4

Figure B-16. Oppenheim Radiation Network from fin surface to surrounding radiation environment

In Figure B-16 the subscript 1 refers to node 1 of the fin and the subscript r refers to the radiation environment surrounding the fin. The radiation environment emissivity, ε_r, and the view factor, F_{1-r}, are both assumed to equal one. If ε_r equals unity, the conductor, $A_r\varepsilon_r/(1 - \varepsilon_r)$, goes to infinity which "shorts out" the conductor and moves the potential, σT_r^4, one node to the left in Figure B-16. The reduced Oppenheim Radiation Network is shown in Figure B-17.

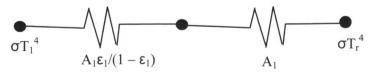

σT_1^4 $A_1\varepsilon_1/(1-\varepsilon_1)$ A_1 σT_r^4

Figure B-17. Reduced Oppenheim Radiation Network from fin surface to surrounding radiation environment

The radiation conductors in Figure B-17 can be combined as follows:

$$\frac{1}{\dfrac{1-\varepsilon_1}{A_1\varepsilon_1} + \dfrac{1}{A_1}} = A_1\varepsilon_1$$

The final reduced Oppenheim Radiation Network and equivalent linearized radiation network are shown in Figure B-18.

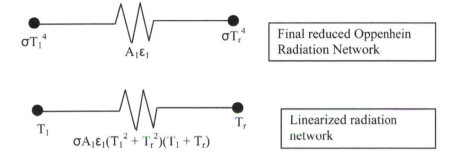

Figure B-18. Radiation Networks

The solution of the fin problem is found by performing an energy balance on each of the four fin nodes and use relaxation to determine the fin temperatures. The steady state energy balance on node 1 is as follows:

$$G_{k\text{-}b}(T_b - T_1) + G_c(T_{ce} - T_1) + G_k(T_2 - T_1) + G_{r\text{-}1}(T_{re} - T_1) = 0$$

Solving for T_1:

$$T_1 = (G_{k\text{-}b}T_b + G_cT_{ce} + G_kT_2 + G_{r\text{-}1}T_{re})/(G_{k\text{-}b} + G_c + G_k + G_{r\text{-}1})$$

The equations for T2, T3, and T4 are as follows:

$$T_2 = (G_kT_1 + G_kT_3 + G_cT_{ce} + G_{r\text{-}2}T_{re})/(2G_k + G_c + G_{r\text{-}2})$$

$$T_3 = (G_kT_2 + G_cT_{ce} + G_kT_4 + G_{r\text{-}3}T_{re})/(2G_k + G_c + G_{r\text{-}3})$$

$$T_4 = (G_kT_3 + G_{c\text{-}e}T_{ce} + G_{r\text{-}4}T_{re})/(G_k + G_{c\text{-}e} + G_{r\text{-}4})$$

Initial temperature values of 1620, 1440, 1260, and 1080 are guessed for nodes 1 through 4 and the equations given above are solved to yield new values for the node temperatures. The new temperature values replace the initial guesses and the process is repeated. At each step, iteration, the radiation conductors must be evaluated since they are functions of the node temperatures. After 117 iterations the temperatures converge, with seven digit accuracy, to the values given below:

T1 = 856.5052°R
T2 = 798.4917°R
T3 = 762.6868°R
T4 = 745.3060°R

NUMERICAL—
COMBINING CONDUCTION AND RADIATION NETWORKS WITH INTERNAL HEAT GENERATION (STEADY STATE)

Examples of internal heat generation are electric heaters attached to surfaces and solar energy absorbed on a surface receiving a solar input. To illustrate this class of heat transfer problem consider the network shown in Figure B-19.

T_r is a boundary node with a fixed temperature of 180°R. The internal heat generation is 3412 Btu/hr, G is 42.08 Btu/(hr-°F), and $\sigma A_1\epsilon_1F_{1\text{-}r}$ equals 1.752E – 08 Btu/(hr-°R⁴). Applying a steady state energy balance at T_b and T_1 and solving for the temperatures we have:

$$T_b = (GT_1 + Q)/G$$

$$T_1 = (GT_b + G_rT_r)/(G + G_r)$$

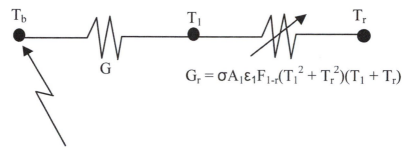

Figure B-19. Steady state problem with conduction, radiation, and internal heat generation.

The results of applying a relaxation to these equations are shown in Table B-10. Inspection of this table shows that the solution is both oscillating and diverging at T_b and T_1!

We have found a situation (steady state conditions with conduction and/or convection, radiation, and internal heat generation) which does not lend itself to steady state relaxation analysis.

Table B-10. Results of applying relaxation to steady state problem with conduction, radiation, and internal heat generation

Iteration	T_b	T_r	T_1	G_r
1	1800	180	900	1.59E+01
2	981.1106509	180	1.35E+03	5.02E+01
3	1435.935588	180	5.45E+02	4.19E+00
4	626.231456	180	1.32E+03	4.69E+01
5	1403.357838	180	3.91E+02	1.85E+00
6	472.1762296	180	1.35E+03	4.99E+01
7	1432.818303	180	3.14E+02	1.13E+00
8	394.750862	180	1.40E+03	5.52E+01
9	1481.126945	180	2.73E+02	8.48E-01
10	354.0287713	180	1.46E+03	6.16E+01
11	1536.521834	180	2.51E+02	7.18E-01
12	331.7135607	180	1.51E+03	6.90E+01
13	1594.85992	180	2.37E+02	6.50E-01
14	318.5921742	180	1.57E+03	7.70E+01
15	1654.456999	180	2.29E+02	6.08E-01

A way around this problem is to use the equation for transient analysis and choose a fictitious capacitance at each node numerically equal to the sum of the conductors connected to the node. For the problem at hand we have the capacitor at node T_b (C_b) = G and the capacitor at node T_1 (C_r) = $(G + G_r)$. The network with the fictitious capacitors is shown in Figure B-20.

Performing transient energy balances on nodes T_b and T_1 we have:

$$G(T_1 - T_b) + Q = (C_b/\Delta t)(T_b' - T_b)$$

$$G(T_b - T_1) + G_r(T_r - T_1) = (C_1/\Delta t)(T_1' - T_1)$$

where, Tb' and T1' are the "new" temperatures after taking the compute time step, Δt.

Solving for Tb' and T1' and substituting G for Cb and $(G + Gr)$ for C1, we have:

$$T_b' = T_b + (\Delta t/G)[G(T_1 - T_b) + Q]$$

$$T_1' = \Delta t/(G + G_r)[G(T_b - T_1) + G_r(T_r - T_1)]$$

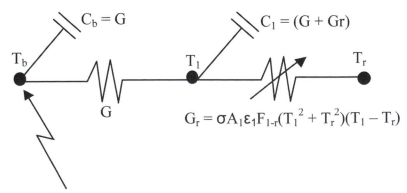

Q, internal heat generation

Figure B-20. Steady state problem with fictitious capacitors C_b and C_1

We let $\Delta t = 1/2*C/$ sum of conductor connected to the node $= \frac{1}{2}$, since we have assigned the value of the capacitor, C, equal to the sum of the conductor connected to the node. This results in the equations:

$$T_b' = T_b + (0.5/G)[G(T_1 - T_b) + Q]$$

$$T_1' = 0.5/(G + G_r)[G(T_b - T_1) + G_r(T_r - T_1)]$$

Using these equations the values for T_b and T_1 converge smoothly in 151 iterations to seven place accuracy.

$$T_b = 746.3061°R \text{ and } T_1 = 665.1955°R$$

There is another approach for solving steady state heat transfer problems with conduction (and/or convection), radiation, and internal heat generation. Instead of linearizing the radiation term we will perform an energy balance leaving in the quartic term. Referring to Figure B-19, the energy balance on node T_b is:

$$Q + G(T_1 - T_b) = 0 \tag{B-39}$$

An energy balance on node T_1 yields:

$$G(T_b - T_1) + \sigma A_1 \varepsilon_1 F_{1\text{-}r}(T_r^4 - T_1^4) = 0 \tag{B-40}$$

Solving Eq. (39) for T_b, we have:

$$T_b = T_1 + Q/G \tag{B-41}$$

Rearranging Eq. (40), we have:

$$T_1^4 + T_1 G/(\sigma A_1 \varepsilon_1 F_{1\text{-}r}) - (GT_b/\sigma A_1 \varepsilon_1 F_{1\text{-}r} + T_r^4) = 0 \tag{B-42}$$

This last equation is a quartic equation which can be solved by the procedure developed by Ferrari[5], which follows.

Rewriting Eq. (B-42) as:

$$T_1^4 + cT_1 + d = 0$$

where:

$$c = G/(\sigma A_1 \varepsilon_1 F_{1-r})$$
$$d = - (GT_b/\sigma A_1 \varepsilon_1 F_{1-r} + T_r^4)$$

$$Q = -d/2$$
$$g = d/3$$
$$f = - (c^2)/16$$
$$G = (f^2 - g^3)^{1/2}$$
$$K = \tfrac{1}{2}[(G - f)^{1/3} - (G + f)^{1/3}]$$
$$S = (2{*}K)^2$$
$$H = K^{1/2}$$
$$V = (S + 2{*}Q)^{1/2}$$
$$N = (2{*}Q)^{1/2}$$

$$T_1 = - H + (V - K)^{1/2}$$

The solution of the problem is found by, first, guessing a temperature for T_b. With this value use the quartic solution, given above, to solve for T_1. Use the value of T_1 and Eq. (B-41) to solve for T_b. Repeat this process until T_1 and T_b converge to constant values. In solving this problem the initially guessed value of T_b was 200°R and after 51 iterations the values of T_1 and T_b converged to 7 place accuracy.

The quartic solution is programmed and given in the Quartic file on the Heat Transfer Helper Disk.

GENERALIZED THERMAL NETWORKS

A generalized thermal network is shown in Figure B-21. The node, x, represents a finite volume or piece of the physical thermal system and the thermal capacitance of a node, Mc_p, is the node mass times the specific heat of the node material. There are two formulations of the generalized equation, a backward differencing, or implicit solution and a forward differencing, or explicit solution. The forward differencing solution allows a step-by-step application of the equation at each node until all nodes are updated and the time step is complete. The backward differencing solution requires the inversion of a matrix involving all the nodes at each time step. When using the forward differencing approach, the size of the time step is limited to a critical time step and if this limit is exceeded, numerical instabilities might occur. This time step restriction does not apply to the backward differencing approach; however, one must be careful and generally not exceed the forward differencing critical time step by more than a factor of 5 or 10 in order to maintain accuracy. Note that steady state solutions are readily analyzed using transient equations by merely continuing the computations until the temperatures no longer change. Thermal capacitances do not apply to steady state problems so any capacitance can be used in applying the transient equations to steady state problems. A convenient approach is to assign a capacitance value numerically equal to the sum of the conductors at the given node. This results in a time step of ½ for the forward differencing equation.

An energy balance on Node x (labeled T_x) in Figure B-21 yields:

$$\sum_{i}^{n} K_i (T_i - T_x) + \sum_{j}^{m} R_j (T_j - T_x) + \sum_{l}^{q} V_l (T_l - T_x) + \sum_{k}^{p} F_k (T_k - T_x) + Q_x =$$

$$(Mc_p)_x \frac{dT_x}{dt} \cong (Mc_p)_x \frac{T_x' - T_x}{\Delta t}$$

(B-43)

K is a conduction conductor = $KA/\Delta x$, k = thermal conductivity, A = area normal to heat flow, Δx = distance of heat flow

V is a convection conductor = hA_{conv}, h = film coefficient, A_{conv} = surface area of node exposed to convection

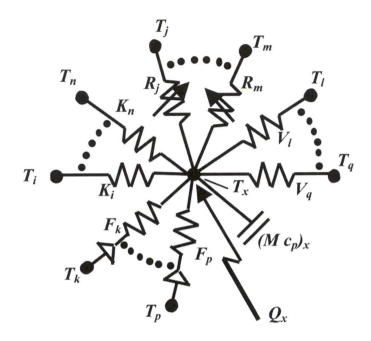

Figure B-21. Generalized thermal network

F is a fluid flow conductor = $m_{dot}c_p$, m_{dot} = mass rate of flow of fluid, c_p = specific heat of fluid

R is a radiation conductor = $\sigma B [(T_j^2 + T_x^2)(T_j + T_x)]$ where B = $A_x\varepsilon_x/(1 - \varepsilon_x)$ or A_xF_{x-j}, σ = Stefan-Boltzmann constant, A_x = radiating area, ε_x = surface emissivity, F_{x-j} = radiation view factor from surface x to j

In equation (B-43) Δt = time step, Q_x = internal heat generation, (for example solar heating or electrical heating), T_x = is the temperature of node x at the beginning of a time step (the old time) T'_x = the temperature of node x after taking a time step (the new time). Solving equation (B-43) for T'_x and choosing all the temperatures on the left side of the energy balance equation at the new time, the backward differencing solution is derived:

$$T_x' = T_x + \frac{\Delta t}{(Mc_p)_x} \left[\sum_i^n K_i\left(T_i' - T_x'\right) + \sum_j^m R_j\left(T_j' - T_x'\right) + \right.$$

$$\left. \sum_l^q V_l\left(T_l' - T_x'\right) + \sum_k^p F_k\left(T_k' - T_x'\right) + Q_x \right] \tag{B-44}$$

If the temperatures on the left side of equation (B-44) are taken at the old time, the forward differencing solution is derived:

$$T_x' = T_x + \frac{\Delta t}{(Mc_p)_x} \left[\sum_i^n K_i\left(T_i - T_x\right) + \sum_j^m R_j\left(T_j - T_x\right) + \right.$$

$$\left. \sum_l^q V_l\left(T_l - T_x\right) + \sum_k^p F_k\left(T_k - T_x\right) + Q_x \right] \tag{B-45}$$

For the forward differencing equation there is a critical time step, as follows:

$$\Delta t_c = \frac{(Mc_p)_x}{\sum_i^n K_i + \sum_j^m R_j + \sum_l^q V_l + \sum_k^p F_k} \tag{B-46}$$

If the critical time step is exceeded the solution may diverge and yield incorrect answers. It is considered good practice to use a compute time step of ½ the critical time step unless fluid flow occurs in the problem and then using the critical time step as the compute time step is preferred.

The simultaneous equations encountered when using the backward differencing approach can be solved using Microsoft Excel. First invert the matrix defined by the variable temperatures at the new time using the Excel function MINVERSE. The inverted matrix is then multiplied by the column matrix defined by the variable temperatures at the old time and the internal heat generation terms. The matrix multiplication is accomplished using the Excel function, MMULT. The Microsoft Excel program used for the matrix manipulations is from Office xp, version 2002. This version of Excel can handle matrices up to a size of 52x52. See section 7 for an example of using Microsoft Excel for solving simultaneous equations.

References

1   Chapman, A.J., Heat Transfer, 2nd Ed., MacMillian, New York, 1960.
2   Schneider, Temperature Response Charts, John Wiley and Sons, Inc., 1963.
3   Arpaci, Conduction Heat Transfer, Addison-Wesley, 1966.
4   Oppenheim, A.K., "Radiation Analysis by the Network Method," Transactions of the ASME, Vol. 78, 1956, p. 725.
5   Seliger, Charles R., Instruments & Control Systems, Vol. 40, February 1967, p 120.

Appendix C

Case Studies from FEMP

INTRODUCTION

Case studies are provided in this appendix to show examples of solar projects that have been implemented in real world applications. The case studies were produced by the Department of Energy Federal Energy Management Program (FEMP), detailing success stories of solar projects at Federal installations. FEMP facilitates the Federal Government's implementation of sound, cost-effective energy management and investment practices to enhance the nation's energy security and environment stewardship.

These case studies can be found on the FEMP website at: http://www1.eere.energy.gov/femp/technologies/renewable_casestudies.html

The case studies included in this appendix are:

- U.S. Marine Corps Base Camp Pendleton Solar Hot Water and Electricity Project

- Phoenix Federal Correctional Institution Solar Hot Water System

- Joshua Tree National Park and Mojave National Preserve Photovoltaic System

- Chickasaw National Recreation Area Solar Hot Water System

- U.S. Navy Pearl Harbor Family Housing Solar Project

U.S. Marine Corps Base Camp Pendleton: Using the Sun for Hot Water and Electricity

Solar hot water collectors and photovoltaic panels on the Camp Pendleton 53 Area training pool are tilted to maximize solar radiation collection.

U.S. Marine Corps Base Camp Pendleton, covering 125,000 acres including 17 miles of Southern-California coastline, is the largest expeditionary training facility on the West Coast. More than 41,500 marines and family members call the base home, which reaches a daytime population of approximately 100,000.

In fiscal year 2007, Camp Pendleton saved energy and money and reduced greenhouse gas (GHG) emissions through solar hot water (SHW) and photovoltaic (PV) arrays. The base implemented two integrated solar thermal/PV systems at its 53 Area and 62 Area training pools. The projects demonstrate Camp Pendleton's continuing commitment to energy conservation while helping meet Federal requirements for on-site renewable energy and solar hot water generation.

System Overview

With a capacity of 500,000 gallons each, the Camp Pendleton training pools provide daily training for Marine Corps personnel year round. The pools typically use natural gas for water heating and electricity for pumps and other mechanical equipment. Camp Pendleton decided to change current practice and take advantage of its abundant solar resources to displace natural gas and electricity consumption.

While solar hot water and photovoltaic technologies have a long history within the Federal sector, Camp Pendleton took a unique, integrated approach. Each pool is equipped with 152 SHW collectors (covering 6,384 square feet) and 108 PV modules (covering 1,485 square feet). The integrated system is supported by a ground-mounted steel structure. Each solar array is tilted to maximize available solar radiation. This approach reduces the system's structural footprint as well as infrastructure and project costs.

Each solar thermal collector is capable of producing 39,400 British thermal units (Btu) of energy each day, resulting in combined annual energy production of 4,371 million Btu (MBtu) for both arrays. As a result, Camp Pendleton eliminates its annual consumption of 54,726 Therms of natural gas for heating the two pools.

Each PV array is rated to generate 31,600 kilo-watt-hours (kWh) of electricity annually, resulting in a combined offset of 63,200 kWh annually.[1] Combined annual electric and natural gas savings for the two training pools is 5,587 MBtu. The Camp Pendleton solar project also reduces annual GHG emissions, including 725,610 pounds of carbon dioxide and 850 pounds of nitrogen oxide.

The Camp Pendleton solar systems were part of a whole-building approach to conserve energy. Other upgrades made to the training pool equipment include variable frequency drives and temperature controls. This equipment allows staff to minimize energy consumption and maximize use of solar resources throughout the day. Temperature controls are tied to a base-wide energy management control system (EMCS) that allows operators to monitor all systems at the training pool from remote loca-

---

[1]Based on manufacturer data and independent testing by the National Solar Rating and Certification Corporation

Integrated solar hot water and photovoltaic panels offset energy consumption for heating and pumping systems at the Camp Pendleton Area 62 training pool.

tions. The result reduces energy costs and the project's payback period.

Project funding originated through a utility energy service contract (UESC) with San Diego Gas & Electric and through Energy Conservation Investment Program (ECIP) funding. The two integrated solar systems cost $1.1 million to construct, but will save the base an estimated $101,600 in electricity and natural gas costs annually. Camp Pendleton offset some of the construction costs through a California Solar Initiative Expected Performance Based Buydown (EPBB) incentive of $90,285. Final payback for the project is less than 10 years.

Technology Overview

Solar hot water systems convert sunlight to thermal energy. A solar collector absorbs the sun's radiation to produce heat. This heat is transferred to water that can then be held in a storage tank for use. There are several types of systems that follow from this basic operation.

Solar water heaters for swimming pools are similar to traditional systems designed for buildings. A collector is mounted to a roof, consisting of a thin, flat, rectangular box that faces the sun. Tubes run through the collector, carrying water that absorbs heat from the collector. The warm water is then circulated to the pool by the pumping and filtration system. The pool itself acts as a storage tank for the hot water.

Photovoltaics, or solar cells, convert sunlight directly into electricity. When sunlight is absorbed by these materials, the solar energy knocks electrons loose from their atoms, allowing the electrons to flow through the material to produce electricity. This process of converting light (photons) to electricity (voltage) is called the photovoltaic effect.

Photovoltaic cells are made from various types of semi-conducting materials. To increase their electrical output, multiple PV cells are assembled to form a panel, which can be assembled to form a larger PV array. Inverters convert direct current (DC) electricity produced by the PV cells to alternating current (AC)—the electric current used to power most appliances and devices.

Project At a Glance	
Federal Facility	Marine Corps Base Camp Pendleton
Pool capacity	500,000 gallons per swimming pool
System Overview	Integrated solar hot water/photovoltaic arrays
SHW Collector Area	6,384 square feet per swimming pool
PV Panel Area	1,485 square feet per swimming pool
Solar Thermal Output	4,371 MBtu annually (combined)
Solar Electricity Output	63,200 kWh annually (combined)
Utility Partner	San Diego Gas & Electric
Year of Completion	2007
Total Cost	$1.1 million
Annual Energy Cost Savings	$101,600
Utility Incentive	$90,285 (California Solar Initiative EPBB)
Payback	10 years

Project Summary

Camp Pendleton's innovative approach to harvesting solar energy significantly increases on-site renewable energy generation while reducing operational costs. The combined projects offset a significant percentage of energy typically consumed in these energy intensive facilities. Utilizing both SHW and PV systems at the training pools ensures that the facilities use less energy and are less dependent on the electric grid and natural gas infrastructure for ongoing operations.

Camp Pendleton completed construction on the 53 Area and 62 Area training pools in 2007. By the end of 2010, Camp Pendleton will have projects in place to add solar thermal and or photovoltaic arrays to all training pools.

Heating Water with Solar Energy Costs Less at the Phoenix Federal Correctional Institution

A large solar thermal system installed at the Phoenix Federal Correctional Institution (FCI) in 1998 heats water for the prison and costs less than buying electricity to heat that water. This renewable energy system provides 70% of the facility's annual hot water needs. The Federal Bureau of Prisons did not incur the up-front cost of this system because it was financed through an Energy Savings Performance Contract (ESPC). The ESPC payments are 10% less than the energy savings so that the prison saves an average of $6,700 per year, providing an immediate payback. Boiler maintenance and hot water service call costs for the facility have also been reduced.

The solar hot water system produces up to 50,000 gallons of hot water daily, enough to meet the needs of 1,250 inmates and staff who use the kitchen, shower, and laundry facilities. Because solar energy is cleaner than conventional electric power, the environment benefits as well. Solar water-heating systems add no carbon dioxide or other emissions to the air around them. This renewable energy system offsets an average annual consumption of 1,000 megawatt-hours (MWh) of electricity and the release of nearly 600 tons of CO_2. For comparison, conventional electricity produced in Arizona emits 1,109 pounds of CO_2 per MWh.

The Federal Bureau of Prisons worked with the Department of Energy (DOE) Federal Energy Management Program (FEMP) and the ESPC contractor, Industrial Solar Technology Corporation (IST), to design and install the system. Under the terms of the 20-year ESPC contract, the prison receives 10% of the total energy savings annually (an average

Parabolic trough concentrator modules at the Phoenix Federal Correctional Institution produce up to 50,000 gallons of hot water daily—enough hot water for kitchen, shower, laundry, and sanitation needs for 1,250 inmates and staff.

of $6,700 per year), and the other 90% goes to amortize the first costs of the system. At the end of the 20-year period, the prison will take over ownership, operation, and maintenance of the solar system and benefit from 100% of the energy savings for the remaining 10 years of the expected service life.

The solar system includes 17,040²ft of parabolic trough concentrating collectors and a 23,000-gallon storage tank located adjacent to the collectors. Parabolic troughs, like other solar water-heating systems, are most cost effective for facilities with relatively constant hot water needs—places such as prisons, hospitals, and barracks. They heat water onsite using the sun's energy, so the facility can reduce the amount of energy purchased from the local utility for water heating.

Highlights

System Capacity—3.4 million Btu/hr (1,000 kW) of heat at 60% peak system efficiency.

Power Production—300 million Btu/month of average delivered heat, offsetting 88,500 kWh/month of electricity consumption to meet 70% of annual need for hot water.

Installation Date—1998

Motivation—Replace large domestic hot-water load heated by electricity with good solar resource 120 parabolic trough concentrator modules.

Size—120 parabolic trough concentrator modules.

Annual savings—$67,000/yr average in electricity costs (90% goes to IST under a 20-year ESPC).

System Details

Components—120 parabolic trough collectors totaling 17.040 ft² (1,584 m²) of collector aperture area; propylene glycol solution circulating fluid; a master field controller and four local sun-tracking controllers.

Storage—Two steel water tanks with membrane liners, totaling 23,000 gallons (87,055 liters)

Loads—13 million Btu/day (4,000 kWh/day) average to heat 30,000-50,000 gallons of water for laundry, kitchen, and

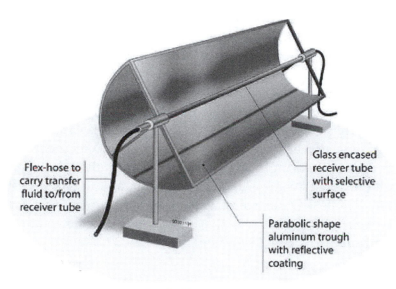

Parabolic trough collector

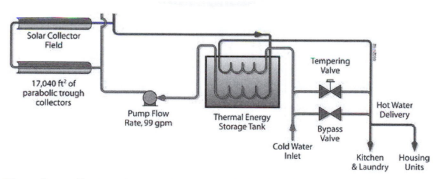

The solar collectors heat a circulating fluid that in turn supplies heat to domestic water storage tanks and end-uses.

other domestic applications.

Supplier/Installer—IST designed, fabricated, installed, and operates the system.

Monitoring—Redundant Btu meters measure delivered hot water; plus a datalogger records solar radiation, wind, ambient temperature, flow rates, and fluid and water temperatures.

Expected Life—30 years.

How The Technology Works

Parabolic trough solar systems convert solar energy to heat. Parabolic trough collectors use mirrored surfaces curved in a linearly extended parabolic shape to concentrate the sun's rays on a pipe running the length of the trough. A mixture of water and antifreeze is pumped through the pipe to pick up the solar energy and then through a heat exchanger to heat potable water. These systems also use single-axis tracking to stay aligned with the sun. Parabolic trough solar systems work well in locations with a high direct-beam solar resource, such as the Southwestern United States. Other solar water heating applications that work well in locations across the country include flat-plate or evacuated-tube collector technology.

Performance

The solar thermal system at Phoenix FCI has been running routinely since March 1999. Under peak solar conditions and when the modules are clean, the solar system delivers up to 3.4 million Btu/hr (1,000 kW) of heat to the energy storage tank at a peak efficiency of 60% of the solar energy incident on the solar collectors. On a sunny day, the solar system delivers up to 50,000 gallons of hot water to the institution, displacing approximately 4,000 kWh of electricity.

On a monthly basis, the system delivers an overall average of 300 million Btu/ month, offsetting 89,000 kWh of electricity consumption and an estimated $5,600 of energy costs. The highest months of energy savings, May 2002 and October 1999, coincide with both the best solar resource (the greatest number of clear sunny days) and the highest hot water demand for prison operations. The lowest months of energy savings, such as October 2001, reflect unusually overcast weather, reduced hot water demand, or partial solar system shut down for maintenance or repairs. To optimize operational efficiency, collectors should be cleaned every 2 to 4 months, depending on weather conditions.

Because calculating the electricity rate is complex and variable, an average blended rate for electricity consumption and

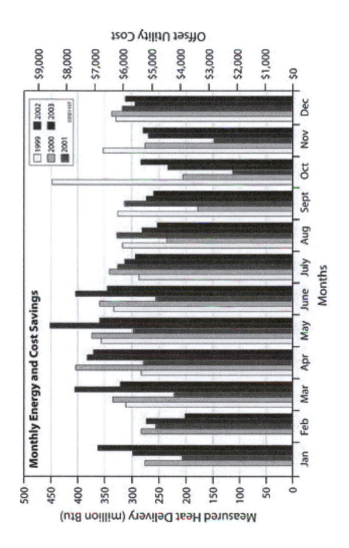

demand is used here to estimate the utility bill savings ($0.065/ kWh). Total annual energy cost savings average $67,000, with 90% going to IST under the ESPC.

Costs

Cost Breakdown for Phoenix FCI Solar Thermal System	
System Cost (total includes: design, hardware, and installation)	$649,000
Per Unit Cost	$38 / ft²
Equivalent Energy Rate	$12/MBtu $0.04/kWh
Annual O&M Cost (rolled into ESPC)	N/A

Life Cycle Cost Analysis for the Phoenix FCI Solar Thermal System		
Study Period: 20 years	Alternative (Electricity utility)	Solar System with Electric Heating Backup
Initial Investment	$ 0	$ 649,000
Recurring Costs (O&M, etc.)	$ 143,419	$ 226,891
Energy Costs	$ 1,528,397	$ 290,465
Total Present Value	$ 1,671,816	$ 1,166,356
Adjusted Internal Rate of Return		6.4 %
Simple Payback Period		8 years
Savings-to-Investment Ratio		1.78

Project Partners and Funding Sources

IST designed, fabricated, and installed the system under an ESPC with the Federal Bureau of Prisons. The ESPC was developed under a Cooperative Research and Development Agreement with the National Renewable Energy Laboratory. Expertise funded by DOE FEMP facilitated the project from feasibility through to performance measurement and verification. The contract term is 20 years, after which ownership of the solar system will revert to the federal government. ABB Energy Capital provided construction and long-term financing to build the system. IST will operate the solar plant over the life of the contract and currently employs the maintenance services of North Canyon Solar.

IST has invested the capital to install and operate the solar thermal system, charging Phoenix FCI a discounted rate for the energy delivered through an ESPC. The project also benefited from a 10% business energy tax credit for purchase of solar equipment and accelerated depreciation of solar energy property investment. DOE FEMP provided cost sharing in the form of technical assistance for this Site-Specific ESPC project, which was the first time a federal agency used an ESPC for a renewable energy technology.

O&M and Emissions Benefits

Operational benefits include maintaining temperatures for domestic hot water (in the past the prison frequently ran out of hot water), reducing electricity peak demand for water heating by more than 200 kW, and reducing maintenance and replacement parts for the offset electric boilers. "We save a lot of money on electric water heater elements, maintenance calls, and repairs," says the facilities manager. "[Plus,] the calls we've gotten from the inmates about cold water have basically gone away." Operation and maintenance savings on the existing boilers are in addition to the reduced utility costs. Furthermore, avoided emissions based on Environmental Protection Agency eGRID 2000 factors for Arizona, amount to 589 tons/yr of CO_2, 2,655 lbs/yr of SO_2, and 2,358 lbs/yr of NO_x.

Federal Renewable Energy Goal

This project is helping the federal government achieve the goal of obtaining 2.5% of electricity from renewable energy by

2005. The Phoenix FCI has one of the largest federal solar thermal systems and one of the earliest renewable energy systems in the U.S. Department of Justice.

A Strong Energy Portfolio for a Strong America

Energy efficiency and clean, renewable energy will mean a stronger economy, a cleaner environment, and greater energy independence for America. Working with a wide array of state, community, industry, and university partners, the U.S. Department of Energy's Office of Energy Efficiency and Renewable Energy invests in a diverse portfolio of energy technologies.

Applications at Other Government Sites

* U.S. Army Fort Sam Houston, San Antonio, Texas: Roof-mounted parabolic troughs provide heat to a pressurized water district-heating loop. Installed June 2003.
* U.S. Army Yuma Proving Ground, Yuma, Arizona: 8,970 million Btu/yr of heat provided for absorption cooling, space heating, and domestic hot water. Installed in 1979 and refurbished in 1986.

Joshua Tree and Mojave Go Solar

Until late 1998, Joshua Tree National Park and Mojave National Preserve in southern California used diesel generators to produce electricity in remote areas. Like many park energy systems, the diesel generators at Joshua Tree's Cottonwood Campground also produced potentially harmful emissions: 120 tons of carbon dioxide, 5,770 pounds of nitrous oxides, 286 pounds of sulfur dioxide, and 218 pounds of suspended particulates every year.

Today, Joshua Tree has cut those emissions dramatically while reducing annual operating costs by an impressive 90%, thanks to a new photovoltaic (PV) system that harnesses the sun's energy to produce clean electric power. Mojave has also had good results. And both parks continue to provide high quality experiences for visitors while preserving our natural resources.

"We've had no trouble," said Harry Carpenter, chief of maintenance at Joshua Tree. "We're happy with the system—it works great."

This project was a combined effort of the Department of the Interior's National Park Service (NPS), Southern California Edison Company's (SCE) Utility Power Group, Sandia National Laboratories, the Department of Energy's Federal Energy Management Program (FEMP), and the National Renewable Energy Laboratory (NREL).

Joshua Tree's Carpenter and Mojave's Dave Paulissen worked with Renew the Parks, a cooperative effort between the NPS Denver Service Center and Sandia's Photovoltaic Design Assistance Center, to find the best renewable energy alternative for the parks. As a result, Joshua Tree and Mojave now have clean, cost-effective PV systems with simple payback periods of only 6.0 years and 8.1 years, respectively.

The Cottonwood Campground Project

Joshua Tree's Cottonwood Campground provides a good example of how these projects worked. It consists of a visitor center, three houses, a duplex, a maintenance building, two small offices, and two pads for campground hosts.

Two 32-kilowatt (kW) diesel generators provided power to the area, consuming about 10,000 gallons (38,000 liters) of diesel fuel per year at a cost of $10,950. Oil replacement every 250 hours added about $1,450 annually for supplies and $4,950 for labor. Annual generator replacement and overhaul costs were about $9,600. The costs of emissions and potential fuel spills brought total estimated annual operating costs to $49,770, or $0.78/kWh.

Sandia recommended replacing the diesel system with a

21-kW PV array, a 250-kWh battery bank, a 30-kW inverter/ battery charger, and a 30-kW backup propane generator. These recommendations, costing about $273,000 to implement fully, had estimated annual operating costs of only about $4,000, with minimal generator use. Replacement batteries ($25,000) would probably be required after about 10 years. Sandia also recommended that the park switch to propane to increase system efficiency and identified several complementary conservation measures. FEMP and NREL were asked to help evaluate project financing options.

A cost-effective, low-emissions photovoltaic array (below) replaced a diesel system at the Joshua Tree National Park Cottonwood Campground area (above).

Innovative Project Financing

FEMP evaluated several alternatives, such as making use of agency funds, energy savings performance contracting, or utility programs. Using appropriated funds would have eliminated financing costs and provided the lowest life-cycle cost, with a discount rate of only 4.1% (in 1996), but funds were not available. Waiting even a few years for appropriations would have cost more in lost savings than the other options. Utility programs offered both technical and financing resources that could be leveraged to implement the project.

The Energy Policy Act of 1992 (Sec. 152) encourages Federal agencies and utilities to design cost-effective demand management and conservation incentive programs. In addition, Federal Acquisition Regulations (FAR) Sec. 41 authorizes agencies to enter into sole-source contracts with regulated utilities under established tariffs. At that time, SCE offered off-grid photovoltaics to Federal facilities through its "Experimental Schedule PVS" tariff—California Public Utilities Commission resolution E-3367.

That tariff allowed SCE to install PV systems, controls, batteries, mounting hardware, and other equipment for a fixed monthly service charge equal to 1.6% of the PV system installed cost (or 19.15% annually) for a term of 15 years. The utility assumed responsibility for system operation and maintenance, so park staff could focus their time and budgets on core mission activities.

Don Zieman, NPS public utility management chief, reviewed procurement and legal issues. Although FAR allows longer than 10-year terms when a tariff provides for it, NPS was more comfortable with a traditional 10-year term. So SCE agreed to 10 years with a 5-year renewal option. A buy-out schedule was also devised in case appropriations became available. And in fact, NPS was able to buy out the contract in the first year following installation.

NPS Solicitor Bill Silver resolved legal questions, and Jack Williams of the NPS Pacific Great Basin Support Systems Office gave the go-ahead to develop the contract. In September 1997, two contracts were signed to provide solar-generated electricity at Joshua Tree and Mojave.

Comparison of Diesel System and Current PV System at Joshua Tree (1996 numbers)		
	Before	**After**
System	2 diesel generators ($32,000)	21 kW PV array, 30-kW propane backup generator $273,000
Fuel Costs/yr	$10,950	$1,100
Emissions/yr	5,770 lb (2,617 kg) NO_X 120 tons (109 metric tons) CO_2 286 lb (130 kg) SO_2 218 lb (99 kg) particulates	382 lb NO_X 7.6 tons (6.9 metric tons) CO_2 0 lb SO_2 0 lb particulates
Total Operating Costs/yr	$49,770	$4,065

*Note: The simple payback period for PV was 6.0 years.

Project Benefits

The PV systems are working well. The Joshua Tree system saves more than $45,000 annually in operating costs. Carbon dioxide emissions are lower by more than 100 tons (90.1 metric tons) per year, NOX emissions are lower by 5,387 lb (2,443 kg), and other emissions have been completely eliminated. Additional benefits include much less generator noise and fewer opportunities for costly spills during fuel transport. Dave Paulissen said that the Mojave system is working so well that another, larger PV system is planned for the Hole-in-the-Wall Fire Center nearby.

The PV systems improve visitors' experiences, reinforce NPS's position as a resource conservation leader, and serve as models for the efficient use of renewable energy. The accomplishments of Renew the Parks projects are being extended under the new Green Energy Parks program, a partnership of the Department of the Interior (through NPS) and the Department of Energy (through FEMP) that will help clear the air in more national parks and recreation areas for years to come.

Showering with the Sun at Chickasaw National Recreation Area

Mark Golnar, a mechanical engineer with the National Park Service (NPS), describes solar water heating as the "perfect heat source" for the comfort stations at Chickasaw National Recreation Area. "The demand for hot water coincides with the availability of sunlight, which makes solar water heaters the obvious choice," he says. "As a bonus, the solar systems are an environmentally sound and cost-effective way to heat water."

Located about 100 miles (161 km) south of Oklahoma City, Oklahoma, on Lake of the Arbuckles, the facility is used primarily in the summer, when solar energy is abundant. The solar water heating systems supply all the hot water for one large comfort station and two small ones at the Buckhorn Campground. (There are no backup systems.) NPS architect James Crockett worked with Golnar to integrate the systems into the design of the new buildings. These solar water heaters are the first of 25 to be installed during the next several years as the site is developed.

Choosing the system

The decision to use solar water heating at the site was the result of a collaborative effort between the National Renewable Energy Laboratory (NREL) Federal Energy Management Program (FEMP) and Solar Process Heat Program in support of NPS. Chickasaw visitors wanted hot showers, and park personnel wanted an alternative to conventional water heaters. The facility had electricity but no propane, and the cost of heating water with electricity was very high.

NREL/FEMP and NPS personnel considered and rejected several solar water heater configurations before deciding on a system design. They determined that high winter stagnation temperatures would damage fluids in a closed-loop antifreeze system, because the campground is rarely used in the winter and there is little demand for hot water. Draindown and re-circulation systems that circulate potable water through the collectors would not work at this site because the hard well water would quickly deposit minerals and obstruct small flow passages. And aesthetic and other site considerations ruled out ground-mounted, tracking parabolic-trough systems.

According to Andy Walker, a member of the NREL/FEMP team, "We concluded that drainback systems, in which the collector heat transfer fluid (in this case water) drains back into the storage tanks when the collector pump turns off, met all the design criteria for this installation. This configuration ensures freeze protection, and even more importantly at Chickasaw, protects the fluid from high stagnation temperatures during the winter months when there is no demand for hot water."

The controller turns the pump on when the collector tem-

Visitors now enjoy hot showers at Chickasaw National Recreation Area, thanks to solar water heating systems such as the one installed on this large comfort station.

perature exceeds the temperature in the storage tank by 20°F (11°C). It turns the pump off when the temperature difference is 4°F (2°C) or less or the temperature in the solar storage tank reaches 180°F (82°C).

The solar heat is transferred from the collector water to domestic water by means of a load-side heat exchanger that consists of coils of copper pipe submerged in the storage tank. The water in the storage tank thus acts as a heat sink from which the incoming domestic water draws stored energy in the form of heat. This strategy increases the reliability of the system, reduces maintenance, and eliminates the need for an external heat exchanger and two pumps.

Because this facility is a recreation area, aesthetics were a primary consideration. As a result, all the collectors are installed unobtrusively on south-facing building roofs. The shading effects of hills, trees, and buildings were not a major concern, because the solar systems collect solar energy mostly in the middle of the day and in the summer, when the sun is overhead.

The solar hot water system on this small comfort station at Chickasaw National Recreation Area supplies all the hot water for the building. There is no backup system.

Chickasaw Project Details

Project Description: Solar water heating systems on new comfort stations

Owner: National Park Service

Location: Chickasaw National Recreation Area, Oklahoma

Architect: James Crockett, National Park Service

Mechanical Engineer: Mark Golnar, National Park Service

Project Supervisor: Brian Lippert, (303) 969-2234

Solar Contractor: SolarMaster Solar Service Inc., supervised by Odes Castor

Chickasaw Facility Manager: Cal Myers

	Small Comfort Stations	Large Comfort Station
Daily hot water use	660 gal (2498 l)/day	1500 gal (5678 l)/day
Temperature	at least 95°F (35°C)	at least 105°F (41°C)
Collector area	194 ft² (18 m²)	484 ft² (45 m²)
Storage volume	500 gal (1893 l)	1000 gal (3785 l)
Load met by solar	9394 kWh	18,194 kWh
Hours water temperature is less than target	345 hr/yr (95°F [35°C])	579 hr/yr (105°F [41°C])
System efficiency	45%	34%
Annual energy savings	9394 kWh/yr	18,194 kWh/yr
Energy saved during 25 years	234,850 kWh	454,850 kWh

Performance

The 194 ft² (18 m²) of collectors on each of the small comfort stations' systems provide 9400 kilowatt-hours (kWh) per year. The installations also include 500 gallons (1893 liters) of hot water storage. The system on the large comfort station has 484 ft² (45 m²) of collectors and provides 18,194 kWh per year and 1000 gallons (3785 liters) of hot water storage.

Because the solar systems are the only sources of hot water at the site, it was important to limit the use of hot water and install conservation devices. The size of the heat exchanger limits instantaneous heat transfer from the storage tank, in effect rationing hot water by limiting the rate at which stored energy is delivered to the load. Tempering valves limit hot water delivery by mixing solar-heated water with cold water to achieve a constant temperature of 105°F (41°C). In addition, the load is minimized by very-low-flow showerheads and 1-minute push button timers on the showers.

NREL's FEMP team did hourly simulations to determine how the systems would perform during an average year. The simulations showed that the small systems will only fail to deliver water hotter than 95°F (35°C) for 345 hours of the year (4% of yearly hours), and the large systems for only 579 hours (7% of yearly hours).

Cal Myers, Facility Manager at Chickasaw, is satisfied with the systems' performance. "Last summer (1997) was the first summer the systems were operational, and I didn't hear any complaints from visitors," he says. "I have experience with maintaining and paying the bills for electric water heaters, and I like having hot water without the bills."

Economics

At Chickasaw, the economics of solar water heating are very attractive. All three systems are cost effective according to the criteria set forth in 10CFR436 of the Code of Federal Regulations (CFR) for Federal facilities. This regulation requires that the life-cycle savings divided by the initial investment (the savings-to-investment ratio) be greater than 1. The Chickasaw systems also fit the definition of cost-effectiveness established in President Clinton's Executive Order 12902 (Energy Efficiency and Water Conservation at Federal Facilities), because they have simple payback periods of fewer than 10 years.

The solar systems at the small comfort stations cost $18,000 each and have a 9-year simple payback, a 6.2% rate of return, and a savings-to-investment ratio of 2.1. The large system cost $24,000, resulting in an 8-year simple payback, a 6.6% rate of return, and a savings-to-investment ratio of 2.4. These figures include the environmental costs of generating electricity based on NPS's assignment of emission costs.

Selling solar

Careful planning and good communication can help ensure the success of renewable energy installations. Solar water heaters are different from conventional water heaters in significant ways. For example, they cost more to buy and are more complex to install and maintain. In addition, maintenance staff are often unfamiliar with the technology and are sometimes resistant to it.

Some Federal managers are also resistant to considering solar water heaters because of past bad experiences. The solar water heating industry experienced failures and bad press in the 1980s, which still linger in some people's minds. And, to make matters worse, it is not uncommon for Federal facilities to have solar water heaters installed on them that have not worked in years—a constant reminder of the deficiencies of early solar equipment.

The good news is that the situation is very different today. After the Federal residential tax credits expired in 1985, many solar companies went out of business. Those that survived continued to refine their products and now sell systems based on mature, reliable, proven technologies. In addition, the Solar Rating and Certification Corporation, a nonprofit national corporation, now develops certification programs and rating standards for solar energy equipment. Most collectors carry 10-year warranties, and systems should last at least 20 years.

In many applications—such as the Chickasaw installations— modern solar water heating systems have economic advantages over conventional water heating systems. A solar water heater's life-cycle costs can be lower than a conventional water heater's, because after the payback period, the solar system continues to produce hot water for only the cost of maintenance.

The big picture

Federal agencies administer more than 31% of the land area in the United States, a large proportion of which is remote and environmentally sensitive. In many of these areas, the need for services is increasing.

In 1991, the NPS identified sustainable design as the cornerstone of an effort to provide services to park visitors without compromising NPS's ability to protect the parks' natural resources. Using renewable energy installations in place of conventional energy technologies has emerged as an important strategy to help carry out this mandate.

Renewable energy technologies are clean, quiet energy sources that help create a more pleasing experience for staff and visitors alike. Solar water heating systems like the ones installed at Chickasaw, for example, consume no fuel, produce no emissions, and—even when economic payoff is small—help agencies fulfill their mandate for responsible stewardship of our national resources.

Leaving a legacy
Protecting the environment is everyone's job, and renewable energy technologies make that possible in tangible and measurable ways. NPS now bases development decisions on life-cycle cost analyses, which include the cost of operating and maintaining installations during their anticipated service life. This is good news for renewable energy technologies, because although they are typically expensive to purchase, they are very competitive on a life-cycle basis.

By using renewable energy technologies to satisfy its mandate to provide services for visitors and protect the park system's natural resources, NPS sets a good example for other Federal agencies and the general public. These technologies offer Federal facility managers the opportunity to comply with Executive Order 12902, meet the cost-effectiveness criteria set forth in 10CFR436 for Federal facilities, and take a step toward leaving their grandchildren a cleaner environment in the bargain.

A Solar Success Story at Moanalua Terrace

Hawaii is a perfect environment for solar water heating," according to Alan Ikeda, a Housing Management Specialist with the Pacific Naval Facility Engineering Command Housing Department in Pearl Harbor, Hawaii. "The sun shines most of the time, we don't have to worry about freezing, the state offers a 35% solar tax credit, and our local utility supports the purchase

COST BREAKDOWN	Small Station	Large Station
Solar System Cost	$18,000	$24,000
Net Present Value of Life-Cycle Cost	$21,300	$28,900
Internal Rate of Return	6.2%	6.6%
Simple Payback Period	9 years	8 years
Discounted Payback Period	10 years	9 years
Savings-to-Investment Ratio	2.1	2.4
NPS Assignment of Emission Costs	**Before 9/97***	**After 9/97***
CO_2	$8/ton ($0.0088/kilogram)	$14/ton ($0.015/kilogram)
SO_2	$0.75/pound ($1.65/kilogram)	$0.85/pound ($1.88/kilogram)
NO_x	$3.40/pound ($7.50/kilogram)	$3.75/pound ($8.33/kilogram)

*The NPS revised its emission costs in September 1997.

Chickasaw Annual Emissions Cost Estimates

SMALL COMFORT STATIONS (each)	Avoided Emissions	Cost of Avoided Emissions (Based on revised emissions costs)
CO_2	12 tons (11,000 kilogram)/yr	$151.00
SO_2	68 pounds (31 kilogram)/yr	$52.00
NO_x	83 pounds (38 kilogram)/yr	$281.00
Annual Value of Avoided Emissions		$484.00
LARGE COMFORT STATION		
CO_2	18 tons (16,000 kilograms)/yr	$227.00
SO_2	101 pounds (46 kilograms)/yr	$78.00
NO_x	123 pounds (56 kilograms)/yr	$416.00
Annual Value of Avoided Emissions		$721.00

Note that this analysis deducts the cost of emissions produced by generating the electricity to run the pumps on the solar systems: (CO_2, $17; SO_2, $6; NO_x, $30; total, $53 for each small station; CO_2, $25; SO_2, $8; NO_x, $45; total, $78 for large station).

25-Year Life-Cycle Cost Analysis

Small Comfort Stations (each)	Basecase	Solar System	Savings
INITIAL INVESTMENT			
Capital Requirements	$3,919	$18,000	-$14,081
FUTURE COSTS			
Recurring Costs*	$9,368	$923	$8,445
Energy-Related Costs**	$24,178	$2,371	$21,807
TOTAL PRESENT VALUE	$37,465	$21,294	$16,172

Large Comfort Station
(one electric heater for basecase, and two solar water heating arrays for solar system)

	Basecase	Solar System	Savings
INITIAL INVESTMENT			
Capital Requirements	$4,875	$24,000	-$19,125
FUTURE COSTS			
Recurring Costs*	$13,965	$1,358	$12,607
Energy-Related Costs**	$36,087	$3,540	$32,547
TOTAL PRESENT VALUE	$54,927	$28,898	$26,029

*Recurring costs, including maintenance costs.
**Energy-related costs, including fuel costs.
Note that basecase is electric water heaters and that these analyses include National Park Service estimates of costs associated with emissions produced by a utility company in the process of generating electricity.

and installation of solar systems with generous rebates."

The Hawaiian Electric Company's (HECO's) $1,500 per unit rebate for solar water heaters installed on new construction helped persuade the Navy to take advantage of Hawaii's solar resource and install solar water heaters on family housing units. At Moanalua Terrace, the Navy had demolished 752 units of family housing, which they are rebuilding in four phases. Designers decided to use the opportunity to give the solar systems a try.

When the 100 homes in Phase I were built, money was not available for solar water heaters. However, Ikeda subsequently secured a $130,000 grant from the U.S. Department of Energy's (DOE's) Federal Energy Management Program (FEMP) to retrofit the Phase I homes with solar systems. In retrofit applications, HECO rebates $800 per unit ($80,000 total) on approved equipment, and Pearl Harbor Family Housing will pay the difference of the estimated $340,000 total cost, or about $130,000.

The 136 units built during Phase II of the Moanalua Terrace project included solar systems in their specifications, so the Navy was able to take advantage of the $1,500 per system HECO rebate for approved solar water heaters in new construction.

The Navy chose direct (open-loop) active systems that circulate potable water through flat-plate collectors coated with a black chrome selective surface. Each system consists of a 4-foot by 8-foot (1.2-m by 2.4-m) collector made by American Energy Technologies, Ltd., and an 80-gallon (302-liter) Rheem tank containing an electric backup element.

Why solar water heating?

Federal agencies, including the military, administer more than 31% of the land area in the United States. In many of these areas, high levels of sunshine make solar technologies a cost-effective energy choice.

Solar water-heating systems offer Federal facility managers a number of advantages compared with conventional water-heating appliances. For one thing, solar water heaters are environmentally benign—they consume no fuel and produce no emissions. Solar water heaters can help Federal facilities comply with 10CFR436 of the Code of Federal Regulations (CFR) for Federal facilities, which, as one of several compliance options, requires that the life-cycle savings divided by the initial investment (the savings-to-investment ratio, or SIR) be greater than one. Solar water systems can also help facility managers comply with Executive Order 12902, which requires increased water conservation and reduced energy consumption in Federal facilities and a significant increase in the use of renewable energy technology to help achieve this goal.

These devices also help reduce Hawaii's dependence on expensive imported oil, and save users money in the bargain. For example, the Pearl Harbor Naval Station has no other energy option for heating domestic hot water except electricity, which currently costs $0.11 per kilowatt- hour. On a life-cycle cost basis, the solar systems are much less expensive than conventional electric water heaters or heat pumps—even without the HECO rebate!

Projects such as Moanalua also benefit solar energy technologies by helping to expand the market for solar water heaters and by fostering familiarity with the technologies among the users. At Moanalua, including the 516 homes now under construction in Phases III and IV and the 100 homes in Phase I that have been retrofitted, this translates to 752 families using solar systems to heat their water.

These projects also offer a public education opportunity. In the case of the Moanalua project, the local utility and the local solar industries association joined forces to present educational programs on the benefits of solar water-heating systems.

Each solar water-heating system at the Navy's Moanalua Terrace housing project offsets about 1.7 tons (1.5 metric tons) of carbon dioxide, 8.2 pounds (3.7 kilograms) of sulphur dioxide, and 11.2 pounds (5 kilograms) of nitrogen oxide every year.

Selling solar

From a facility manager's perspective, solar water heaters do present challenges compared with electric water heaters. They are more expensive to purchase, more complex to install and maintain, and—perhaps most important—residents and maintenance personnel are often unfamiliar with the technology.

In addition, some Federal managers are resistant to considering solar water heaters because of past bad experiences. The solar water-heating industry experienced failures and bad press in the 1980s, but the situation is very different today. The solar companies that survived the demise of the Federal residential tax credits in 1985 continued to refine their products. They now sell systems based on mature, reliable, proven technologies. In addition, the Solar Rating and Certification Corporation, a nonprofit national corporation, develops certification programs and rating standards for solar energy equipment. Most collectors carry 10-year warranties, and systems should last at least 20 years.

In many applications—including the Moanalua installations—modern solar water-heating systems have economic advantages over conventional water-heating systems. At Moanalua, the solar water heaters' life-cycle costs are lower than those of either a conventional electric water heater or a heat pump, and, after the payback period, the solar system continues to produce hot water for only the cost of maintenance. Solar water heaters also come in different configurations that can be matched to specific applications to maximize efficiency and cost effectiveness. Solar water heaters' many advantages make any resistance worth overcoming.

Economics

The Navy had originally planned to use high efficiency electric water heaters in its new and renovated family housing, but decided on solar water heaters because of HECO's Residential New Construction Rebate and Residential Efficient Water Heating Rebate Programs. As the Navy's research revealed, solar water heating is a particularly attractive choice in Hawaii, because of that state's unique circumstances.

Almost 90% of Hawaii's power comes from imported oil, and Hawaii utilities are required by law to pass on fuel price increases within 30 days. If oil prices increased suddenly, the shock to Hawaiian consumers would be almost immediate. Hawaii burns about 2.1 million barrels of oil per year (more than 5,000 barrels per day) to heat domestic water, and solar water heaters can significantly reduce this environmental and economic burden. For example, the 50,000 solar water-heating systems currently operating in Hawaii avoid the burning of more than 350,000 barrels of oil each year. A typical Hawaiian family of four can avoid the burning of seven barrels of oil per year by installing a solar water-heating system.

The criteria set forth in 10CFR436 require that the SIR be greater than one. The definition of cost-effectiveness established in President Clinton's Executive Order 12902 is that

At Moanalua Terrace, the Navy's analysis revealed that solar water heaters were more cost effective than either high-efficiency electric water heaters or heat pumps.

systems have simple payback periods of less than 10 years.

According to the Navy's Ikeda, "An analysis of the economics of solar water heating at Moanalua shows that solar water heating is a better investment than either heat pumps or electric water heaters with or without the rebate over a 20-year period." Although the solar systems are more expensive to buy and install initially, they use far less energy to operate during a 20-year period. Even without the rebate, the Moanalua systems have a simple payback period of about 10 years and an SIR of 1.4. With the HECO rebate of $1,500 per solar water-heating system, the simple payback is about six years and the SIR is 2.5.

Throughout a 20-year life cycle heat pumps are more cost effective than electric water heaters, but far less cost effective than solar water heaters. According to the Navy's research, even with an HECO rebate of $725 per unit, heat pumps have a simple payback of more than 40 years and an SIR of only 0.67.

Performance

To ensure eligibility for the HECO rebate, solar water heaters must provide 90% of the average annual water heating load, perform consistently throughout the life of the system, and have an estimated useful life of at least 15 years. Both the Navy and the residents are satisfied with the performance of the solar systems at Moanalua.

Initially, residents complained that they didn't have enough hot water because the utility only allowed the backup element to come on for one hour per day. However, since the timers were adjusted to allow the backup elements to operate for a maximum of 2 hours per day, there have been no complaints.

Leaving a legacy

Protecting the environment is everyone's job, and renewable energy technologies make that possible with reliable, cost-effective equipment. Some businesses and government agencies are beginning to base development decisions on life-cycle cost analyses, which include the cost of operating and maintaining installations throughout their anticipated service life. This is good news for renewable energy technologies, because although they are typically expensive to purchase, they require minimal maintenance, no fuel, and produce no emissions, which makes them a good long-term investment.

At Moanalua, solar water heaters are the most economical water-heating option, even without the utility rebate. If we include the estimated costs of avoided emissions in the mix, a good investment looks even better. Based on the U.S. Environmental Protection Agency's Emission Factors for calculating emissions offsets, each solar system at Moanalua offsets about 1.7 tons (1.5 metric tons) of carbon dioxide, 8.2 pounds (3.7 kilograms) of sulphur dioxide, and 11.2 pounds (5 kilograms) of nitrogen oxides every year. According to the National Park Service's assignment of costs to those emissions offsets, this amounts to an additional annual dollar savings of $72.77 per unit. Multiplied by the 136 units in Phase II of the Moanalua project, the savings is almost $10,000 per year.

By using solar water heaters to provide hot water for family housing, the Navy demonstrates environmental awareness and good financial management. Other Federal agencies can realize the same benefits by following the Navy's good example at Moanalua Terrace.

References

1   Chapman, A.J., *Heat Transfer*, 2nd Ed., MacMillian, New York, 1960.
2   Schneider, *Temperature Response Charts*, John Wiley and Sons, Inc., 1963.
3   Arpaci, *Conduction Heat Transfer*, Addison-Wesley, 1966.
4   Oppenheim, A.K., "Radiation Analysis by the Network Method," Transactions of the ASME, Vol. 78, 1956, p. 725.
5   Seliger, Charles R., *Instruments & Control Systems*, Vol. 40, February 1967, p 120.

20-Year Life-Cycle Cost Analysis

Including HECO rebate:

	Basecase (electric)	Solar System (90% solar fraction)	Savings (per unit)
INITIAL INVESTMENT			
Incremental Cost/Unit	$400	$2,310	-$1,910
FUTURE COSTS			
Recurring Costs*	$659	$741	-$82
Energy Costs**	$5,352	$539	+$4,813
TOTAL PRESENT VALUE	$6,411	$3,590	$2,821

*Includes maintenance and capital replacement costs discounted over 20 years using the National Institute of Standards and Technology's (NIST's) 1998 discount rate of 4.1%.
**Based on the U.S. Department of Energy's (DOE's) projected energy prices for electricity in DOE Region 4.

Savings-to-investment ratio (SIR) = 2.5
Simple payback = 6 years
Discounted payback = 7 years

Note that this analysis includes a $1,500 per unit rebate from the Hawaiian Electric Company and does not include the costs associated with emissions produced by the utility company to generate electricity. Analysis performed using NIST's Building Life-Cycle Cost (BLCC) software tool.

MOANALUA PHASE II ANNUAL AVOIDED EMISSIONS COST ESTIMATES*

	Avoided Emissions**	Cost of Avoided Emissions***
CO_2	1.7 tons (1543 kilograms)	$23.80
SO_2	8.2 pounds (3.7 kilograms)	$6.97
NO_x	11.2 pounds (5.1 kilograms)	$42.00
Annual Value of Avoided Emissions		$72.77

*Per unit, compared with an electric water heater.
**Based on the U.S. Environmental Protection Agency's Regional Emission Factors.
***Based on the National Park Service (NPS) assignment of costs to emissions, revised 9/97.

NPS Assignment of Emission Costs

CO_2	$14/ton ($0.015/kilogram)
SO_2	$0.85/pound ($1.88/kilogram)
NO_x	$3.75/pound ($8.33/kilogram)

Note that this analysis does not include the cost of emissions produced by generating the electricity to run the pumps on the solar systems.

Moanalua Project Details

Project Description: Solar water-heating systems installed on 136 units of new Navy housing

Location: Oahu, Hawaii

Design: R&R Solar Supply, Honolulu, Hawaii

Installation: Dorvin D. Leis Company, Inc.

Collector: AE-32 from American Energy Technologies, Ltd.

Pump: PACO 006B

Controls: Heliotrope DTT94

Tank: Rheem 81VTCR80-1 80-gallon (302-liter) top connect

Index

A

absorber plate/fin 50, 52, 55, 58-60, 102, 119-121

absorptivity 27, 30-32, 49, 95

altitude angle-solar 15, 64, 123

apparent solar irradiation 12, 15

atmospheric extinction coefficient 13

axis tracking 63-68, 79, 85, 86, 90, 104, 172

azimuth angle 10, 18, 63-65, 71, 81, 82, 86, 94, 103, 104, 123

C

Carnot engine 87-91, 111,117

clearness numbers 17-18

concentrating collector 49, 170

concentrating solar power 79

conduction 19-21, 27, 39, 48, 83, 119, 120, 124, 127, 131, 132, 138, 149, 150, 151, 154, 155, 157, 159, 162

convection 19, 22-27, 39, 54, 60, 83, 95, 96, 107, 119, 120, 124, 127, 131, 132, 138, 149, 150, 151, 155, 157, 159

D

declination 10-13, 15, 64, 101

direct normal incident radiation 15-17, 61

E

efficiency-collector 52-57, 61, 63, 66

emissivity 27-32, 36, 37, 46, 49, 54, 103, 144, 145, 151, 152, 160

F

film coefficient 23-25, 40, 42, 54, 60, 74, 75, 95, 96, 104, 119-121, 126, 133-135, 138, 139, 141, 151, 159

flat plate collector 49, 51, 53, 56, 190

free convection 22, 25, 95, 96

fluid flow 22, 23, 26, 39, 42-44, 50, 54, 133-143, 160, 161

G

Grashof number 24, 101

H

heat transfer coefficient 73-77, 104

heliostat 1, 79-85, 104

hot water-solar 1, 49, 69-72, 104, 108, 163, 164-167, 168-176, 181-187, 188-197

I

incident absorbed solar radiation 63, 66, 67

infrared emissivity 30-32, 49, 103

isothermal 35

L

laminar flow 23, 25

M

monochromatic emissivity 28-31

monochromatic absorptivity 31-32

N

Nusselt number 24, 25

O
Oppenheim Radiation Network 35-
   38, 144, 152, 153

P
passive solar 93-98, 100, 105
Percent of Possible Sunshine 14, 18,
   62, 70, 71, 73
photovoltaic systems 1, 79, 85, 93,
   104, 163-168, 177-179
Planck's equation 28
power tower 79-85
Prandtl number 24, 25

R
radiation 5, 19, 26-39, 45, 54, 59, 80,
   83, 95, 101, 144-157, 160
radiosity 35
reflectivity 27, 54, 103
refraction 6
Reynolds number 23-26

S
solar coefficients 13, 14
solar flux 32
solar radiation 1, 5, 9, 12, 13, 15-19,

45, 49, 50, 52, 54, 57, 61-67, 86,
   90, 95, 98, 103, 104, 164, 165, 171
space heating-solar 73-77, 176
specific heat 24, 42, 45, 95, 111, 114,
   124, 127, 131, 133, 135, 139-141,
   159, 160
Stefan-Boltzmann constant 27-33, 46,
   144, 149, 152, 160
Stirling engine 87-92, 105, 115
System Loss Factor 70, 72

T
thermal conductivity 1, 20, 21, 24,
   60, 74, 95, 120, 121, 124, 126, 127,
   149, 151, 159
thermal network 39, 42, 43, 74, 75,
   97, 130-132, 136, 137, 141, 143,
   144, 158, 160
transmissivity 27, 50, 54, 103
Trombe wall 95-98
turbulent flow 23, 24, 26

V
vacuum tube collector 49, 50
view factors 15, 33-37, 46, 144, 146,
   151, 152, 160